T0176428

3D Object Processing

3D Object Processing
Compression, Indexing and Watermarking

Edited by

Jean-Luc Dugelay
Institut Eurécom, France

Atilla Baskurt
INSA Lyon, CNRS LIRIS, France

Mohamed Daoudi
TELECOM Lille1, CNRS LIFL, France

John Wiley & Sons, Ltd

Other Wiley Editorial Offices

John Wiley & Sons Inc., 111 River Street, Hoboken, NJ 07030, USA

Jossey-Bass, 989 Market Street, San Francisco, CA 94103-1741, USA

Wiley-VCH Verlag GmbH, Boschstr. 12, D-69469 Weinheim, Germany

John Wiley & Sons Australia Ltd, 42 McDougall Street, Milton, Queensland 4064, Australia

John Wiley & Sons (Asia) Pte Ltd, 2 Clementi Loop #02-01, Jin Xing Distripark, Singapore 129809

John Wiley & Sons Canada Ltd, 6045 Freemont Blvd, Mississauga, Ontario, L5R 4J3, Canada

Wiley also publishes its books in a variety of electronic formats. Some content that appears
in print may not be available in electronic books.

Library of Congress Cataloging-in-Publication Data

3D object processing : compression, indexing, and watermarking / edited by Jean-Luc Dugelay,
Atilla Baskurt, Mohamed Daoudi.
    p. cm.
    Includes bibliographical references and index.
    ISBN 978-0-470-06542-6 (cloth)
    1. Image processing – Digital techniques. 2. Three-dimensional imaging.
3. Digital images – Watermarking. 4. Image compression. I. Dugelay, Jean-Luc.
II. Baskurt, Atilla, 1960- III. Daoudi, Mohamed, Ph. D.
    TA1637.A144 2008
    621.36'7 – dc22

                  2008003728

British Library Cataloguing in Publication Data
A catalogue record for this book is available from the British Library

ISBN 978-0-470-06542-6 (HB)

Typeset in 10.5/13 Times by Laserwords Private Limited, Chennai, India
Printed and bound in Great Britain by TJ International, Padstow, Cornwall

Contents

About the Contributors

Tarik Filali Ansary received a BE degree from Ecole Marocaine des Sciences de l'Ingénieur, Morocco, in 2002, and an MS degree from the University of Sciences and Technologies of Lille (USTL), France, in 2003, both in Computer Science. He received a PhD in 2006 from the Institut National des Télécommunications, France, and at USTL with the FOX-MIIRE Research Group at the LIFL (UMR USTL/CNRS 8022), a research laboratory in the computer science field of the USTL. His principal research interests include 3D content-based retrieval, multimedia indexing, and retrieval from content. Email: tarik.filali@gmail.com.

Atilla Baskurt was born in Ankara, Turkey, in 1960. He received his BS degree in 1984, MS in 1985 and PhD in 1989, all in Electrical Engineering from INSA of Lyon, France. From 1989 to 1998, he was Maître de Conférences at INSA of Lyon. Since 1998, he has been Professor in Electrical and Computer Engineering, first at the University Claude Bernard of Lyon, and now at INSA of Lyon. Since 2003, he has been the Chair of the Telecommunication Department. Professor Baskurt leads the research activities of two teams at the LIRIS Research Laboratory: the IMAGINE team and the M2DisCo team. These teams work on image and 3D data analysis and segmentation for image compression, image retrieval, shape detection and identification. His technical research and experience include digital image processing, 2D–3D data analysis, compression, retrieval and watermarking, especially for multimedia applications. He is Chargé de Mission on Information and Communication Technologies (ICT) at the French Research Ministry. Email: atilla.baskurt@insa-lyon.fr/abaskurt@liris.cnrs.fr.

Jihane Bennour received her engineering degree from CPE (Lyon, France) in 2004 and in parallel she obtained an MS degree in Image Signal Vision from the University of St-Étienne (France). She worked for six months in 2004 at INRIA (Sophia Antipolis, France) with the ORION team on the generation of 3D animations from scenario models. In October 2004 she started her PhD thesis in the Image and Video Group for Multimedia Communications and Applications at the Institut Eurécom under the guidance of Professor Jean-Luc Dugelay. Her research interest is 3D object watermarking. She was formerly involved in the FR-RNRT SEMANTIC-3D project and the European Ecrypt project. Email: jihane.bennour@eurecom.fr.

Stefano Berretti received a doctoral degree in Electronics Engineering and a PhD in Computer Engineering and Communications from the University of Florence, Italy, in 1997 and 2001, respectively. In 2000, he received a post-laurea degree in Multimedia Content Design from the Master in Multimedia of the University of Florence. Since 2002, he has been Assistant Professor at the University of Florence, where he teaches operating systems. Stefano Berretti also teaches database systems for the Master's degree in Multimedia of the University of Florence. His current research interests are focused mainly on content modeling, retrieval and indexing of images and 3D object databases. Email: berretti@dsi.unifi.it.

Mohamed Daoudi is Professor of Computer Science in Institut TELECOM; TELECOM-Lille1 and LIFL (UMR USTL/CNRS 8022). He received his PhD in Computer Engineering from the University of Lille 1 (USTL), France, in 1993 and Habilitation à Diriger des Recherches (HDR) from the University of Littoral, France, in 2000. He was the founder and head of the MIIRE Research Group of LIFL from 2000 to 2004. His research interests include pattern recognition, image processing, invariant representation of images and shapes, 3D analysis and retrieval and, more recently, 3D face recognition. He has published more than 80 papers in refereed journals and proceedings of international conferences. He served as a Program Committee member for the International Conference on Pattern Recognition (ICPR) in 2004 and the International Conference on Multimedia and Expo (ICME) in 2004 and 2005. He is a frequent reviewer for *IEEE Transactions on Pattern Analysis and Machine Intelligence* and for *Pattern Recognition Letters*. His research has been funded by ANR, RNRT, European Commission grants. Email: mohamed.daoudi@telecom-lille1.eu.

Alberto Del Bimbo is Professor of Computer Engineering and the Director of the Master's degree in Multimedia of the University of Florence, Italy. He was the Director of the Department of Systems and Information, from 1997 to 2000 and the Deputy Rector for Research and Innovation Transfer of the University of Florence from 2000 to 2006. From 2007 he has been the President of the Foundation for Research and Innovation, promoted by the University of Florence. His scientific interests are pattern recognition, image databases, human–computer interaction and multimedia applications. He has published over 230 papers in some of the most distinguished scientific journals and presented at international conferences, and is the author of the monograph *Visual Information Retrieval*, on content-based retrieval from image and video databases, published by Morgan Kaufmann in 1999. From 1996 to 2000, he was the President of the IAPR Italian Chapter, and, from 1998 to 2000, Member at Large of the IEEE Publication Board. He was the General Chair of IAPR ICIAP'97, the International Conference on Image Analysis and Processing, IEEE ICMCS'99, the International Conference on Multimedia Computing and Systems, AVIVDiLib'05, the International Workshop on Audio-Visual Content and Information Visualization, and VMDL07, the International Workshop on Visual and Multimedia Digital Libraries, and was on the board of many other primary scientific conferences. He is Associate Editor of *Pattern Recognition, Journal of Visual Languages and Computing, Multimedia Tools and Applications, Pattern Analysis and Applications* and *International Journal of Image and Video Processing*, and was Associate Editor of *IEEE Transactions on Multimedia* and *IEEE Transactions on Pattern Analysis and Machine Intelligence*. Email: delbimbo@dsi.unifi.it.

Jean-Luc Dugelay (PhD 1992, Member of the IEEE 1994, Senior Member 2002) joined the Institut Eurécom (Sophia Antipolis) in 1992, where he is currently a professor in charge of image and video research and teaching activities in the Department of Multimedia Communications. His research interests are in the area of multimedia signal processing and communications, including security imaging (i.e. watermarking and biometrics) and facial image analysis. He contributed to the first book on watermarking (*Information hiding techniques for steganography and digital watermarking*, Artech House, 1999). He is an author or co-author of more than 150 papers that have appeared in journals or proceedings, 3 book chapters and 3 international patents. He has given several tutorials on digital watermarking (co-authored with F. Petitcolas from Microsoft Research, Cambridge) at major conferences (ACM Multimedia, October 2000, Los Angeles, and Second IEEE Pacific-Rim Conference on Multimedia, October 2001, Beijing). He has been an invited speaker and/or member of the Program Committee of several scientific conferences and workshops related to digital watermarking. He was technical co-chair and organizer of the Fourth Workshop on Multimedia Signal Processing, Cannes, October 2001. His group is involved in several national and European projects related to digital watermarking (RNRT Aquamars and Semantic-3D, IST Certimark). Professor Dugelay served as associate editor for several journals and is currently the Editor in Chief of the *EURASIP Journal on Image and Video Processing*. He is currently a member of several Technical Committees of the IEEE Signal Processing Society. Email. jld@eurecom.fr.

Florent Dupont received his BS and MS degrees in 1990, and his PhD in 1994, from INSA in Lyons, France. Since 1998, he has been Associate Professor. In 2001, he joined the LIRIS Laboratory (UMR 5205 CNRS) in the University Claude Bernard of Lyon. His research interest mainly concerns 3D digital image processing, 3D compression and discrete geometry. He is involved in the M2Disco research team which aims at proposing new combinatorial, discrete and multiresolution models to analyze and manage various types of data such as images, 3D volumes and 3D meshes. Email: Florent.Dupont@liris.cnrs.fr.

Emmanuel Garcia obtained his MSc from Ecole Polytechnique, France, in 1998, and an MSc from Ecole Nationale Supérieure des Télécommunications, France, in 2000, and his Ph.D. from Université de Nice–Sophia Antipolis, France, in 2004, on the subject 'Texture-based watermarking of 3D video objects'. Since 2005 he has been employed as an image processing expert at Infoterra France. His research interests concern image processing and computer vision, now applied to photogrammetry, cartography and city modeling. Email: emmanuel.garcia@infoterra.fr.

Guillaume Lavoué has an Engineering Degree in Electronic, Signal Processing and Computer Science from CPE-Lyon (France), an MSc in Image Processing from the University Jean Monnet of St-Étienne (France) and a PhD in Computer Science from the University Claude Bernard of Lyon (France). From February to April 2006, he was a postdoctoral fellow at the Signal Processing Institute (EPFL) in Switzerland. Since September 2006, he has been Associate Professor at the French engineering university INSA of Lyon. He is involved in the M2Disco team from the LIRIS Laboratory (UMR 5205 CNRS). His research interests focus on 3D model analysis and processing, including compression, watermarking, geometric modeling and 2D/3D recognition. Email: glavoue@liris.cnrs.fr.

Nikos Nikolaidis received his Diploma of Electrical Engineering in 1991 and PhD in Electrical Engineering in 1997, both from the Aristotle University of Thessaloniki, Greece. From 1992 to 1996 he served as a teaching assistant in the Departments of Electrical Engineering and Informatics at the same university. From 1998 to 2002 he was a postdoctoral researcher and teaching assistant at the Department of Informatics, Aristotle University of Thessaloniki. He is currently Assistant Professor in the same department. Dr Nikolaidis is the co-author of the book *3-D Image Processing Algorithms* (John Wiley & Sons, Ltd, 2000). He has co-authored 7 book chapters, 26 journal papers and 87 conference papers. He currently serves as Associate Editor on the *International Journal of Innovative Computing Information and Control*, the *International Journal of Innovative Computing Information and Control Express Letters* and the EURASIP *Journal on Image and Video Processing*. His research interests include computer graphics, image and video processing and analysis, copyright protection of multimedia and 3D image processing. Dr Nikolaidis is a Member of the IEEE and EURASIP. Email: nikolaid@aiia.csd.auth.gr.

Pietro Pala graduated in Electronics Engineering from the University of Florence, Italy, in 1994. From the same university, he received a PhD in Information and Telecommunications Engineering in 1997. Presently, he is Associate Professor of Computer Engineering at the University of Florence where he teaches database management systems and image and video processing and analysis. He also teaches multimedia systems for the Master's degree in Multimedia Content Design of the University of Florence. His main scientific and research interests include pattern recognition, image and video databases and multimedia information retrieval. Email: pala@dsi.unifi.it.

Julien Tierny received an MSc degree (summa cum laude) in Computer Science from Lille University (USTL) along with an engineering degree from TELECOM-Lille1 in 2005. He is currently a PhD candidate within the Computer Science Laboratory of the University of Lille (LIFL, UMR USTL/CNRS 8022). He is also a teaching assistant in the Computer Science Department of the University of Lille. His research interests include shape modeling, shape similarity estimation, geometry processing and their applications. Email: julien.tierny@telecom-lille1.eu.

Jean-Philippe Vandeborre received his MS degree in 1997 and PhD in Computer Science in 2002 from the University of Sciences and Technologies of Lille (USTL), France. Currently, he is Associate Professor of the

Institut TELECOM; TELECOM Lille 1 (a graduate engineering school in information technology and communications), France, in the Computer Science and Network Department where he teaches TCP/IP networks, Unix systems and computer graphics. He is also a member of the LIFL (UMR USTL/CNRS 8022), a research laboratory in the computer science field of the University of Sciences and Technologies of Lille, linked to the CNRS (Centre National de la Recherche Scientifique). His current research interests are mainly focused on 3D model analysis, and include multimedia indexing and retrieval from content and their applications. He is involved in French national research projects as well as in the European Network of Excellence DELOS. Email: jean-philippe.vandeborre@telecom-lille1.eu.

Introduction

Scientific and technological advances in the fields of image processing, telecommunications and computer graphics during the last decade have contributed to the emergence of new multimedia, especially three-dimensional (3D) digital data. Nowadays, the processing, transmission and visualization of 3D objects are a part of possible and realistic functionalities over the Internet. Confirmed 3D processing techniques exist and a large scientific community works hard on open problems and new challenges, including progressive transmission, fast access to huge 3D databases or content security management.

The emergence of 3D media is also directly related to the emergence of the 3D acquisition technologies. Indeed, recent advances in 3D scanner acquisition and 3D graphics rendering technologies boost the creation of 3D model archives for several application domains. These include archaeology, cultural heritage, computer-assisted design (CAD), medicine, 3D face recognition, videogames or bioinformatics.

Finally, this growing 3D activity is possible thanks to the development of hardware and software for both professionals (especially 3D modeling and tools for creation and manipulation) and for end-users (3D graphic accelerated hardware, Web3D, high-quality PDAs (Personal Data Assistants), new generation of cellular/mobile phones to visualize 3D models interactively).

Why is our Community Particularly interested in these New Media?

A 3D object is more complex to handle than other multimedia data, such as audio signals, images or videos. Indeed, only a unique and simple two-dimensional (2D) grid representation is associated to a 2D image. All the

3D Object Processing: Compression, Indexing and Watermarking   J.-L. Dugelay, A. Baskurt & M. Daoudi
© 2008 John Wiley & Sons, Ltd

2D acquisition devices generate this same representation (digital cameras, scanners or 2D medical systems). In consequence, any image processing function can be applied directly to all these images issuing from different sensors. This does not mean that the function will lead to satisfactory results for a universal heterogeneous 2D database; this function has to be adapted to the characteristics of different classes of 2D images, for instance.

Unfortunately (for the users) and fortunately (for the scientists), there exist different 3D representations for a 3D object. An object can be represented on a 3D grid like a digital image, or in 3D Euclidean space. In the latter case, the object can be expressed by a single equation (like algebraic implicit surfaces), by a set of facets representing its boundary surface or by a set of mathematical surfaces. The reader can easily imagine the numerous open problems related to these different representations.

The sources generating 3D data will not produce the same 3D representation for the same object. Furthermore, the different application domains have their usual types of representation. Either one needs to have a universal processing function which works with all the representations (this is inconceivable), or one needs to know how to transform a representation to the other one: this is quite complex and often constitutes open problems.

The Focus of the Book

For this book, we deliberately limit our ambition and focus on three important and complementary topics concerning 3D data: compression, indexing and retrieval, and watermarking. This choice also defines clearly the scenario of services and uses we consider: namely, *secure transmission, sharing and finding of similar 3D objects on networks*. All the four chapters are written with this scenario in mind.

Whatever the problem to resolve (compression to guarantee a rate/distortion/channel capacity constraint on networks; indexing to facilitate 3D retrieval; watermarking for copyright protection and content security and integrity improvement), we need to manipulate the same 3D representation models commonly used in computer graphics and 3D modeling. This is why it seemed very important to us to detail these models in the first chapter of this book. Before going on to the functional sections (compression, indexing or watermarking), the reader needs to visit this theoretical introduction in order to understand the characteristics, advantages and limitations of the

3D representation schemes presented in three categories: polygonal meshes, surface-based models (parametric, implicit and subdivision surfaces) and volumetric models (primitive based (superquadric, hyperquadric), voxel representation, CSG).

Chapter 2 presents the 3D coding/decoding schemes based on the different categories of 3D representation models previously presented in Chapter 1. These schemes correspond to efficient tools to reduce the size of the 3D data and the transmission time for low-bandwidth applications while minimizing the whole distortion and preserving the psychovisual quality of the 3D objects. Whatever the scheme, two main parts constitute challenging open problems: finding a concise/synthetic representation of the 3D model, even if it implies some loss of precision or some errors regarding the original object; and detecting and removing the redundancy of this model. This chapter presents two classes of schemes: those based on the well-known 3D representation, the polygonal meshes; and those which are based on new representation models instead of classical polygonal meshes (especially, more compact models like NURBS, implicit surface or subdivision surface).

Chapter 3 is devoted to the indexing and retrieval of 3D shapes: how to access and exploit 3D digital collections using visual information as queries. In recent years, many content-based image retrieval (CBIR) systems have been proposed in the literature for 2D images and videos. However, when the information is intrinsically 3D, as in computed-assisted design, a simple generalization of existing CBIR from 2D to 3D is inconceivable. One needs to reconsider the problem from scratch. This chapter presents the very recent 3D techniques for indexing, comparing and retrieving 3D objects. The approaches can be classified into three classes: view-based approaches (inspired from human perception of 3D structures that is based mainly on 2D profiles), structural approaches (based on topological descriptions of 3D shapes using for example skeleton or a set of elementary geometric objects) and full-3D approaches (based on features computed directly on a 3D space).

Finally, Chapter 4 presents the 3D watermarking methods. The past decade has seen the emergence of 3D media in the industrial, medical and entertainment domains. Therefore, the intellectual property protection and authentication problems of these media have attracted more attention in these years. It has to be noted that the community has just begun to work on this very challenging topic. The techniques are not as abundant and ripe as is the case for other media such as images, audio and video. They have to be tested in terms of robustness on large 3D databases in order to be validated by the whole community. This chapter presents the ones with the most potential for hiding information in a 3D object without altering the visual quality and for recovering this information at any time, even if the 3D object was altered by one or more non-destructive attacks, either malicious or not.

Without further ado, we invite the reader to explore and enjoy this book.

1

Basic Background in 3D Object Processing

Guillaume Lavoué

1.1 3D Representation and Models

This chapter details the different 3D representation models, commonly used in computer graphics and 3D modeling. We distinguish three main representation schemes: polygonal meshes; surface-based models, such as parametric, implicit and subdivision surfaces; and volumetric models including primitive-based models (superquadric, hyperquadric), voxel representation and constructive solid geometry. For each of these 3D representation schemes, we analyze the advantages and drawbacks in terms of modeling, regarding different applications.

The point-based representation, which has been recently introduced in computer graphics, is not within the scope of this book. Recent reviews can be found in Kobbelt and Botsch (2004) and Alexa *et al.* (2004).

1.1.1 Basic Notions of 3D Object Representation

Now 3D objects are more complex to handle than other multimedia data, such as audio signals or 2D images, since there exist many different representations for such objects.

3D Object Processing: Compression, Indexing and Watermarking   J.-L. Dugelay, A. Baskurt & M. Daoudi
© 2008 John Wiley & Sons, Ltd

For instance, a 2D image has a unique and rather simple representation: a 2D grid ($n \times n$) composed of n^2 elements (named *pixels*) each containing a color value or a gray level. The different devices and techniques which produce digital images (digital cameras, scanners, etc.) all provide the same representation.

For 3D models there are different kinds of representation: an object can be represented on a 3D grid like a digital image, or in 3D Euclidean space. In the latter case, the object can be expressed by a single equation (like algebraic implicit surfaces), by a set of facets representing its boundary surface or by a set of mathematical surfaces.

The main difficulties with a 3D object are that:

- The different sources of 3D data (tomography, laser scanning) do not produce the same representations.
- The different applications (computer-aided design, medical) do not consider the same representations.
- Changing from one representation to another is quite complex and often constitutes open problems.

Figure 1.1 illustrates several representations of the 3D object *Bunny*.

From a real-world object, a laser scanner produces a set of points in 3D space, each defined by its coordinates x, y and z. Figure 1.1(a) illustrates a cloud of points representing the object *Bunny*. This point-based representation, provided by scanners, is not efficient for computing physical or geometrical properties or displaying on a screen. Indeed there are no neighborhood relations between 3D points, neither is there any surface or volumetric information. Therefore these representations are often converted to polygonal meshes and particularly *triangular* meshes (see Figure 1.1(b)). With this model, the object is represented by its boundary surface which is composed of a set of planar faces (often triangles). More precisely, a polygonal mesh contains a set of 3D points (the *vertices*) which are linked by *edges* to form a set of polygonal *facets*. Many *surface reconstruction* techniques exist to produce a triangular mesh starting from a cloud of points issuing from a 3D scanner.

Polygonal meshes can represent open or closed surfaces from arbitrary topology, with a precision that depends on the number of vertices and facets. Intersection, collision detection or rendering algorithms are simple and fast with this model, since manipulating planar faces (and particularly triangles) is also simple (linear algebra). This rapidity is particularly useful

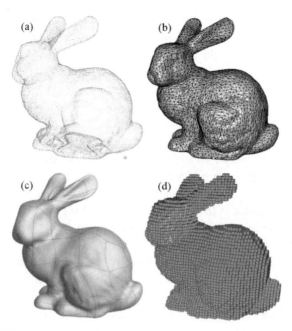

Figure 1.1 Different representations of *Bunny*: (a) a cloud of points, (b) a triangular mesh, (c) a set of parametric surfaces, (d) a set of voxels

for videogames. These benefits make this model the most widespread representation for 3D objects.

However, polygonal meshes have some limitations. This model is intrinsically *discrete* since the number of vertices and facets depends on the expected precision, thus a high precision can lead to a huge amount of data. Moreover, the definition of the shape is very local and thus applying a global deformation or manually creating a shape facet by facet is quite difficult.

The need for a 3D model adapted for modeling and conception has led to the appearance of parametric surfaces. This family of 3D surfaces (including Bézier, B-spline and NURBS surfaces) is particularly used for computer-aided design (CAD), manufacturing (CAM) and engineering (CAE). This model enables mathematically exact surfaces to be defined contrary to polygonal meshes which only represent an approximation. Since these surfaces are defined on a parametric domain, they cannot represent a shape with arbitrary topology, hence a 3D object is often represented by a patchwork of parametric surfaces. For instance, *Bunny* from Figure 1.1 is modeled with 153 bicubic B-spline patches.

These surfaces have very strong tangential and curvature continuity properties that make them quite useful for design. Moreover, they are mathematically complete and allow modeling of a large variety of shapes. Since they are defined only by a few sets of control points, instead of a dense set of vertices, they are also much more compact than polygonal meshes. The main drawback is that they are more complex to manipulate than a set of triangles.

Polygonal meshes and parametric surfaces are boundary representations (BREP) since they model an object only by its boundary. Several domains (especially medical imaging) need the interior data of a 3D object and thus consider a volumetric model and particularly the *discrete* representation. This model does not represent the 3D object in Euclidean space, but in a 3D grid similar to the 2D image representation. Each element of the grid is a voxel (short for volumetric pixel). Thus the object is represented by the set of voxels constituting its volume. A voxel can contain a boolean value (*in* the object or *outside* the object) or other information like local densities. Medical devices (RMI for instance) often produce such volumetric data to describe the interior of an organ. Figure 1.1(d) illustrates the *voxelized Bunny*, in a $50 \times 50 \times 50$ grid.

1.1.2 Polygonal Meshes

A 3D polygonal mesh is defined by a set of plane polygons (see Figure 1.2). Thus, such a model contains three kinds of *elements*: *vertices*, *edges* and *faces* (the most used are triangles). Additional attributes can be attached to the vertices, such as normal vectors, color or texture information. A polygonal mesh consists of two kinds of information: the *geometry* and the *connectivity*. The geometry describes the position of the vertices in 3D space and the connectivity describes how to connect these positions, i.e. the relationship between the mesh elements. These relations specify, for each face, the edges and vertices which compose it, and for each vertex, the incident edges and faces. The *valence* of a vertex is the number of its incident edges and the *degree* of a face is its number of edges (see Figure 1.2). A polygonal mesh is called *manifold* if each of its edges belongs to one or two faces (one in the case of a border). Manifold meshes present strong geometrical properties, and thus are considered in almost all the existing mesh compression methods. A manifold mesh is associated with its Euler–Poincaré characteristic χ:

$$\chi = v - e + f \tag{1.1}$$

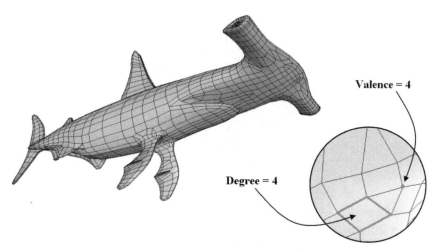

Figure 1.2 Polygonal mesh example (2560 vertices, 2562 faces, 5120 edges), illustrating the *valence* of a vertex and the *degree* of a face

with v, e and f, respectively, the numbers of vertices, edges and faces of the mesh. The Euler–Poincaré characteristic is related to the genus of the corresponding surface, according to the following equation:

$$g = \frac{2c - b - \chi}{2} \tag{1.2}$$

where c is the number of connected components and b the number of borders. The genus of a surface describes its topological complexity; it corresponds to the maximum number of closed curves, without common points, which can be traced inside this surface without disconnecting it (i.e. the complement to these curves remains connected). Basically the genus is the number of *handles* of the mesh; a sphere and a torus are respectively of genuses 0 and 1. Figure 1.3 illustrates a genus-65 model.

The polygonal mesh model is by far the most popular in computer graphics and 3D modeling. Its algebraic simplicity (linear algebra) largely facilitates algorithms of intersection, collision detection or rendering. Moreover, polygonal meshes have a high capacity of description; indeed any complex object of arbitrary topology can be modeled by a polygonal mesh, provided there is enough memory space available. Finally, many techniques generate a polygonal mesh starting from other geometrical models (parametric surfaces, implicit surfaces, discrete models, etc.). This model representation largely

Figure 1.3   Example of a genus-65 3D mesh (8830 vertices, 17919 faces). Courtesy of SensAble technologies by the AIM@SHAPE Shape Repository

dominates the various 3D interchange formats, such as VRML and MPEG-4 (2002).

Nevertheless this model has several limitations: it is a *discrete* representation, because it depends on the targeted scale. Indeed, a polygonal mesh only possesses a C^0 continuity, and thus cannot represent exactly a smooth surface, only an approximation whose precision (linked to the number of polygons) depends on the targeted scale and application (rendering, collision detection, etc.). Consequently, this model can become quite heavy in terms of the amount of data if the expected precision is high, or if the object to be represented is complex.

The standard mesh format (used in the VRML standard, for instance) is as follows. The geometry is represented by a list of coordinates indexed over the vertices and the connectivity is described by the list of faces, each of them being represented by a cyclic list of indices of its incident vertices. Figure 1.4 illustrates this standard representation. The file contains the coordinates of the vertices V_0, V_1, V_2, V_3, V_4, V_5 (the *geometry*) and for each face F_0, F_1, F_2, F_3, the indices of the corresponding vertices (the *connectivity*).

Although this coding method offers the advantage of simplicity, encoded data are very redundant since vertices and edges are referred to several times. In the binary format, for a triangular mesh, each of the three coordinates of each vertex is basically encoded by a *float* (4 bytes) and for each face, each of the three vertex indices is encoded by an *integer* (4 bytes). If n_s is the number of vertices and n_f the number of faces, then the size of the file

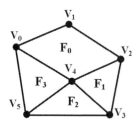

Geometry	
V_0	$x_0\,y_0\,z_0$
V_1	$x_1\,y_1\,z_1$
V_2	$x_2\,y_2\,z_2$
V_3	$x_3\,y_3\,z_3$
V_4	$x_4\,y_4\,z_4$
V_5	$x_5\,y_5\,z_5$

Connectivity	
F_0	0 1 2 4
F_1	4 2 3
F_2	4 3 5
F_3	0 4 5

Figure 1.4   Example of standard representation for a simple mesh, containing six vertices and four faces

is $(n_f + n_s) \times 4 \times 3$ bytes. For a regular mesh, for which $n_f \approx 2 \times n_s$, we have to encode $3 \times 4 \times 3 = 36$ bytes (or 288 bits) per vertex.

1.1.3   Surface-based Models

In this class of representation, a 3D object is represented by its boundary, defined by one or more pieces of a surface. We detail here the three main representation schemes comprising this family: parametric surfaces, implicit surfaces and subdivision surfaces. Polygonal meshes also belong to this class of representation, but due to its predominance in computer graphics, we have devoted a whole section to its description.

Parametric surfaces

Description

The parametric representation of a 3D surface defines each point $S(\eta, \mu)$ of the surface by the following equation:

$$S(\eta, \mu) = \begin{bmatrix} f_x(\eta, \mu) \\ f_y(\eta, \mu) \\ f_z(\eta, \mu) \end{bmatrix} \qquad (1.3)$$

η and μ are the two *parametric* variables, and f_x, f_y and f_z are functions from $\mathbb{R}^2$ in $\mathbb{R}$.

Parametric models represent a large family of surfaces. The most common are the Bézier surfaces and their extensions, B-spline and NURBS

(Non-Uniform Rational B-Spline) surfaces, which are particularly popular in CAD and CAM.

Pierre Bézier introduced parametric surfaces in the field of CAD in 1972 (Bézier 1972). Bézier curves and surfaces are characterized by a set of control points, which defines an $n \times m$ grid. The linear combination of these control points, following the two directions η and μ, provides all the surface points $S(\eta, \mu)$:

$$S(\eta, \mu) = \sum_{i=1}^{n} \sum_{j=1}^{m} P_{i,j} B_i^n(\eta) B_j^m(\mu) \qquad (1.4)$$

where $P_{i,j}$ are the control points, and $B_i^k(t) = C_k^i t^i (1-t)^{k-i}$ the Bernstein polynomials (with $t \in [0, 1]$), which represent the blending functions. Figure 1.5 illustrates a bicubic Bézier surface, defined on a 4×4 grid of control points. Although they have strong continuity properties, these surfaces have two major drawbacks:

- No local control is possible, i.e. the displacement of a control point involves modifications on the whole surface.
- The polynomial degree increases with the number of control points. Thus it is computationally expensive to handle complex control polyhedra.

The necessity to eliminate these drawbacks has led to the creation of a new model: B-spline curves and surfaces. The formulation is quite similar to Bézier surfaces, with different blending functions. Bernstein polynomials are

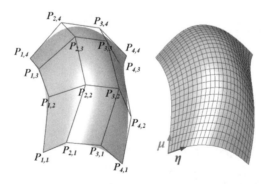

Figure 1.5 Bicubic Bézier surface (16 control points) and isoparametric curves

replaced by the B-spline functions. Considering k and l ($k \leq n$ and $l \leq m$), the respective degrees of the surface in the two directions η and μ, the formulation is:

$$S(\eta, \mu) = \sum_{i=1}^{n} \sum_{j=1}^{m} P_{i,j} N_i^k(\eta) N_j^l(\mu) \qquad (1.5)$$

$N_i^k(t)$ is the B-spline basis function, of degree k, recursively defined by:

$$N_i^k(t) = \frac{(t - t_i) N_i^{k-1}(t)}{t_{i+k} - t_i} + \frac{(t_{i+k+1} - t) N_{i+1}^{k-1}(t)}{t_{i+k+1} - t_{i+1}} \qquad (1.6)$$

with:

$$N_i^0(t) = 1 \text{ if } t \in [t_i, t_{i+1}[\qquad (1.7)$$

$$N_i^0(t) = 0 \text{ otherwise} \qquad (1.8)$$

A B-spline surface is thus a continuous, piecewise polynomial surface defined like the union of surface patches of fixed degrees (k and l) connected according to the blending B-spline functions. Vectors $t_0, t_1, \ldots, t_{n+k+1}$ and $t_0, t_1, \ldots, t_{m+l+1}$ are called the node vectors and form monotonous increasing floating values lists which represent the positions of the beginning and end of the compact supports of the blending functions. In other words, they define which proportions of each patch are present in the B-spline surface. If the nodes are uniformly spaced, the B-spline is said to be *uniform*. Figure 1.6 details this mechanism for 2D curves: a B-spline segment (analogous to a surface *patch*) is illustrated on the left while a B-spline curve made up of three B-spline segments is shown on the right.

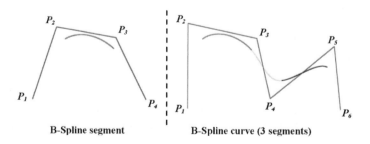

B-Spline segment   B-Spline curve (3 segments)

Figure 1.6   B-spline curves

B-spline curves and surfaces have many benefits, compared with Bézier ones. They permit local control: the displacement of a control point involves modifications on a delimited region of the surface (depending on degrees k and l). Moreover, the complexity and the size of the control polyhedron do not influence the degree and thus the calculation complexity of the polynomials. Finally k and l control the order of B-spline functions, therefore for the same set of control points, the resulting surface can more or less match the control polyhedron.

The next logical step is the NURBS model. These surfaces are a generalization of B-splines. A weight w_{ij} is associated with each control point. This coefficient is a tension parameter: increasing the weight of a control point pulls the surface toward that control point. NURBS models enable the definition of not only natural quadrics (such as spheres, cylinders, etc.) but also surfaces having sharp edges. For more details on parametric surfaces, interested readers are invited to consult the book by Farin (1993).

Evaluation

The representational power of parametric surfaces is really strong, especially in the case of NURBS. Further, this model has been included in the VRML (dedicated to the Internet) and MPEG-4 (2002) standards and in various 3D CAD exchange formats (IGES, STEP, etc.).

The benefits of these surfaces are numerous: they are mathematically complete and therefore easy to sample or to digitize into voxels, triangles, etc. Moreover, they do not depend on a scale factor (unlike discrete models like meshes or voxels). Lastly, they have strong continuity properties (particularly useful for design), and are much more compact in terms of amount of data than the polygonal mesh model.

However, parametric models are not as widely used as polygonal meshes, for various reasons: they remain quite complex to manipulate, and the algorithms of collision detection and intersection are quite difficult to implement. Finding intersection points between two NURBS (or B-spline) surfaces amounts to solving a complex mathematical problem (Krishnan and Manocha 1997) whereas this operation is commonplace for polygonal meshes. Also, these surfaces are defined on parametric grids, which makes them difficult to fit to an object with arbitrary topology. Thus, a whole object is often composed of several NURBS patches, associated with trimming curves which do not yet have a real standardization and which increase the volume of data.

Implicit surfaces

Description

Consider an implicit surface S as the zero set of a function f from $\mathbb{R}^3$ in $\mathbb{R}$. The set of points $P = (x, y, z)$ of the surface S defined by the function f verifies:

$$f(x, y, z) = 0 \tag{1.9}$$

Common shapes are often defined in such manner, for instance a straight line $ax + by + cz + d = 0$ or a sphere $x^2 + y^2 + z^2 - r^2 = 0$.

Thanks to this formulation, the main property of this model is to give directly the relation between the surface and every point of the 3D space. Hence, this model defines not only a surface but also a solid; indeed the sign of the function f specifies a whole partition of the 3D space. A point $P = (x, y, z)$ is such that:

- $f(x, y, z) > 0$ is outside the implicit surface;
- $f(x, y, z) < 0$ is inside the implicit surface;
- $f(x, y, z) = 0$ is on the implicit surface.

Implicit surfaces can be classified into two principal categories: *algebraic* and *non-algebraic*. A surface is algebraic if f is a polynomial. Figure 1.7 depicts such a surface. Superquadrics and hyperquadrics belong to this class, and are described in section 1.1.4.

Since algebraic surfaces are quite complex to manipulate, another formulation (non-algebraic) for implicit surfaces has been developed. The main idea

Figure 1.7 Implicit surface associated to the algebraic equation $x^4 - 5x^2 + y^4 - 5y^2 + z^4 - 5z^2 + 11.8 = 0$

is to represent an object by a set of particles. Each particle is represented by its center (a point of the 3D space) and a potential function $\Phi(r)$ decreasing with the distance r to the center. If the scene is composed with n particles, the value of the global potential function Φ^g for a point P of the 3D space is defined by:

$$\Phi^g = \sum_{i=1}^{n} \Phi_i(r_i) \qquad (1.10)$$

with r_i the distance from P to the center of the ith particle. The summation of the potential functions is a sort of generalization of union operators and enables the blending of different particles. Corresponding isosurfaces (surfaces corresponding to a constant value for Φ^g) then allow modeling of arbitrary (even complex) shapes.

The first author to propose such a model was Blinn (1982), who introduced the *blob* objects defined by a Gaussian potential function $\Phi(r) = be^{-ar^2}$. Figure 1.8 illustrates the blending of two *blobs* associated with different distances.

The main drawback of Blinn's model is its *global* definition; indeed the potential function Φ is not bounded. Thus other potential functions have been introduced (Murakami and Ichihara 1986; Nishimura *et al.* 1985; Wyvill *et al.* 1986) which permit bounding of the *blob* influence in order to speed up calculations.

This model was improved and generalized by replacing the center of the particles by some more complex primitives like segments or faces, in order to create *skeleton-based* implicit surfaces. The potential field value generated by an element of the skeleton, for a point P of the 3D space, is then defined by a potential function $\Phi(r)$, with r the distance from P to the

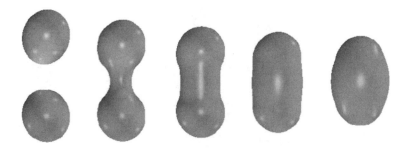

Figure 1.8 Unions of two *blobs* at different distances

skeleton (Bloomenthal and Wyvill 1990). However, this modeling technique can produce artifacts or discontinuities, particularly for a non-convex skeleton (when the radius of curvature of the skeleton is smaller than the radius of influence of the potential function, for instance). Thus a new class of skeleton-based implicit surfaces was introduced, based on the principle of convolution (Bloomenthal and Shoemake 1991). These surfaces are now defined by using integrals of the potential field along the skeleton elements. Numerous authors are developing more and more complex and effective implicit models: Wyvill *et al.* (1999) introduced the *BlobTree* which organizes the elements of the skeleton in a hierarchical graph structure, in which nodes represent blending operations, boolean operations or deformations. This model was extended by Barbier *et al.* (2005) who introduced levels of detail. In the same way, Angelidis and Cani (2002) and Hornus *et al.* (2003) presented a multiresolution implicit model, based on convolution along a skeleton represented by subdivision curves.

Evaluation

Through their different representations (algebraic, particles, skeletons), implicit surfaces offer real modeling power from the more intuitive (*blobs*, skeletons) to the more *mathematical* (algebraic surfaces). Arbitrary topologies can also be modeled (provided that there is no boundary or sharp edge). Moreover, these surfaces make the algorithms of intersection, belonging or blending particularly simple. Deformation and animation algorithms are also easily facilitated, hence implicit surfaces are especially used in the medical field, for the modeling of body parts. Further this model is particularly compact. In the case of a sphere, for example, the implicit formulation is quite simple: $(x - c_x)^2 + (y - c_y)^2 + (z - c_z)^2 - r^2 = 0$.

However, sampling and distance calculations are quite difficult to handle. Contrary to the parametric models, points of the implicit surfaces are not directly calculable. Thus, triangulate surfaces require quite complex algorithms like the *marching cubes* (Lorensen and Cline 1987) or the more recent *marching triangles* (Akkouche and Galin 2001). In spite of a relatively high representational power, due particularly to the recent non-algebraic implicit modeling techniques (Angelidis and Cani 2002; Barbier *et al.* 2005; Bloomenthal and Shoemake 1991; Bloomenthal and Wyvill 1990; Hornus *et al.* 2003; Muraki 1991), implicit surfaces are especially adapted for organic modeling, and cannot represent mechanical or CAD objects because of the

difficulty of representing planes or sharp edges. Concerning algebraic surfaces, there is no possible local control, or local deformations.

Subdivision surfaces

Description

The basic idea of subdivision is to define a smooth shape from a coarse polyhedron by repeatedly and infinitely adding new vertices and edges according
to certain subdivision rules. Figure 1.9 illustrates this subdivision mechanism for a 2D curve. Figure 1.9(a) shows four points connected by segments
(the control polygon), while Figure 1.9(b) shows the polygonal curve after a
refinement step: three points have been added and the old points have been
displaced. By repeating this refinement several times, the polygonal curve
takes on the appearance of a smooth curve. Continuity properties of the limit
curve depend on the subdivision rules. Similarly, an example of a subdivision surface is presented in Figure 1.10. At each iteration, each quadrangle
is divided into four, new points are thus inserted and the old ones are displaced. After several subdivision iterations, the surface seems smooth (see
Figure 1.10(d)).

The main difficulty is to find subdivision rules which lead to certain good
properties such as calculation simplicity, local control, continuity and pleas-
ant visual aspect. In the case of the subdivision curve presented in Figure 1.9,
the polygonal curve is firstly linearly subdivided (i.e. new points are added
in the middle of the segments), and then each point P_i is replaced by a linear
combination of itself and its direct neighbors P_{i-1} and P_{i+1}, following the
smoothing mask:

$$P'_i = \frac{1}{4}(P'_{i-1} + 2P_i + P'_{i+1}) \tag{1.11}$$

With these rules, the limit curve corresponds to a uniform cubic B-spline
(see Figure 1.9).

(a)   (b)   (c)   (d)

Figure 1.9 Example of a subdivision curve: (a) control polygon, (b, c) two refinement
steps, (d) limit curve

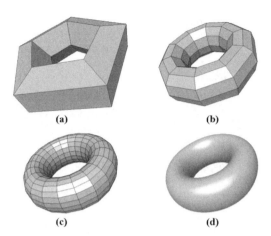

Figure 1.10 Example of a subdivision surface: (a) control polyhedron, (b, c) two refinement steps, (d) limit surface

Doo and Sabin (1978) and Catmull and Clark (1978) (see Figure 1.10) were the first authors to introduce subdivision surface rules. Their schemes respectively generalized biquadratic and bicubic tensor product B-splines (Farin 1993). Today, many subdivision schemes have been developed, based on quadrilateral (Kobbelt 1996; Peters and Reif 1997) or triangular meshes (Loop 1987; Zorin *et al.* 1996). Recently, Stam and Loop (2003) and Schaefer and Warren (2005) have introduced subdivision schemes for mixed quad/ triangle polyhedra. Moreover, special rules have been introduced by Hoppe *et al.* (1994) to handle sharp edges (i.e. to preserve the sharpness of a given control edge). A subdivision scheme can be described by:

- **A topological component**: Every subdivision scheme changes the connectivity of the polyhedron, by adding/removing vertices or flipping edges. Then we can classify them into two main categories: *primal* schemes which do not remove old vertices, and *dual* schemes which remove them.
- **A geometric component**: The displacement of the vertices can be seen as a smoothing operation on the input polyhedron. Considering this operation, subdivision schemes can also be split into two main classes: *interpolating* schemes which do not modify the position of the old vertices, and *non-interpolating* schemes which change the position of both old and new vertices.

Table 1.1   Classification of principal subdivision schemes

Primal	Triangles	Quadrangles	*Dual* (quadrangles)
Non-interpolating	*Loop* (C^2) (Loop 1987)	*Catmull–Clark* (C^2) (Catmull and Clark 1978)	*Doo–Sabin* (C^1) (Doo and Sabin 1978) *Midedge* (C^1) (Peters and Reif 1997)
Interpolating	*Butterfly* (C^1) (Zorin *et al.* 1996)	*Kobbelt* (C^1) (Kobbelt 1996)	

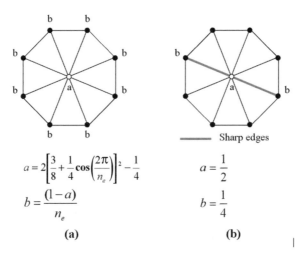

$$a = 2\left[\frac{3}{8} + \frac{1}{4}\cos\left(\frac{2\pi}{n_e}\right)\right]^2 - \frac{1}{4}$$

$$b = \frac{(1-a)}{n_e}$$

$$a = \frac{1}{2}$$

$$b = \frac{1}{4}$$

(a)   **(b)**

Figure 1.11   Smoothing masks for Loop subdivision rules: (a) standard vertex, (b) *sharp* vertex

Hence, we can classify existing subdivision schemes. This classification is detailed in Table 1.1.

A subdivision scheme is generally described by smoothing masks. Figure 1.11 illustrates the different masks associated with the Loop (Loop 1987) subdivision rules which apply on triangular meshes. We consider the subdivision scheme as being composed of two distinct steps: a linear subdivision (a vertex is added at the middle of each edge), and then a smoothing (each vertex is replaced by a linear combination of itself and its direct neighbors). Smoothing coefficients are presented in Figure 1.11(a) with n_e the valence of the considered vertex. Special rules were introduced by Hoppe *et al.* (1994) for handling sharp and boundary edges. Figure 1.11(b) illustrates the smoothing coefficients for vertices shared by two sharp edges.

For more details about subdivision surfaces, the reader may refer to the book by Warren and Weimer (2002).

Evaluation

Subdivision surfaces offer many benefits. Firstly they can be generated from arbitrary meshes (arbitrary topology), which implies no need for trimming curves (which are necessary for NURBS surfaces, defined on parametric rectangles). Secondly, the majority of existing subdivision schemes are really easy to implement and linear in complexity. Further, they can be generated at any level of detail, by iterative subdivision of the control mesh, according to the terminal capacity for instance. Local control is possible (which is not the case for algebraic implicit surfaces) with the possibility of creating and preserving sharp features. Most of the existing schemes produce at least C^1 continuous (except around sharp edges, of course) limit surfaces. Lastly, it is an extremely compact model since a smooth surface is represented by a coarse control mesh (thus no need to encode node vectors or trimming curves as for the NURBS). For all these reasons, subdivision surfaces are now widely used in computer graphics and 3D imaging, and particularly in the fields of animation, reconstruction and CAD, and have been integrated into the MPEG-4 standard (MPEG-4 2002).

One of the main drawbacks of subdivision is the lack of parametric formulations. Thus they cannot be directly evaluated at any points. Moreover, the control mesh has to be adapted to the considered rules (based on quads or triangles). Also, for certain subdivision schemes, the continuity properties are not as strong as for parametric surfaces, particularly around *extraordinary* points (vertices of the mesh having a non-regular valence).

1.1.4 Volumetric Representation

Primitive-based models

These models decompose a complex 3D object into a set of simple ones, called primitives, which are then combined using different operations. These primitives are generally arranged in a graph that allows a structural modeling of the object, particularly useful for indexing and recognition purposes.

Several types of primitives exist: generalized cylinders (Binford 1971; Ponce *et al.* 1989; Zeroug and Nevatia 1996), geons (Biederman 1985; Nguyen and Levine 1996), superquadrics (Barr 1981), supershapes (Gielis *et al.* 2003), hyperquadrics (Han *et al.* 1993) and other implicit polynomial models (Keren *et al.* 1994). These models can be classified in two main categories: *quantitative* models having a real representational power (superquadrics, hyperquadrics, implicit polynomials) and *qualitative* models which have a symbolic and structural representation purpose (geons, generalized cylinders). The objective of this chapter is to present true modeling techniques, thus we will only detail quantitative models and more particularly superquadric models and their extensions of supershapes and hyperquadrics, implicit polynomial models having already been detailed in Section 1.1.3.

Superquadrics

Superquadrics were introduced by Barr (1981) in the early 1980s. They are an extension of quadrics: two additional parameters were introduced, ε_1 and ε_2, which control the latitudinal and longitudinal curvatures. These primitives have the specificity to possess both implicit and parametric formulations. The most commonly used ones are superellipsoids, defined as follows.

Parametric formulation:

$$S(\eta, \mu) = \begin{bmatrix} a_1 \cos^{\varepsilon_1}(\eta) \cos^{\varepsilon_2}(\mu) \\ a_2 \cos^{\varepsilon_1}(\eta) \sin^{\varepsilon_2}(\mu) \\ a_3 \sin^{\varepsilon_1}(\eta) \end{bmatrix} ; \ -\frac{\pi}{2} \le \eta \le \frac{\pi}{2}, -\pi \le \mu \le \pi \quad (1.12)$$

Implicit formulation:

$$F(x, y, z) = 1 - \left[\left(\frac{x}{a_1} \right)^{\frac{2}{\varepsilon_2}} + \left(\frac{y}{a_2} \right)^{\frac{2}{\varepsilon_2}} \right]^{\frac{\varepsilon_2}{\varepsilon_1}} + \left(\frac{z}{a_3} \right)^{\frac{2}{\varepsilon_1}} \quad (1.13)$$

Parameters a_1, a_2, a_3 control the size of the superellipsoid while ε_1 and ε_2 control its curvature. Figure 1.12 illustrates some examples of superellipsoids associated to different ε_1 and ε_2 values. Hence, with only five parameters, this model yields the representation of a quite broad range of shapes such as cubes, cylinders, octahedra or ellipsoids. Barr (1981) has also defined two other classes of superquadric, namely supertoroid and superhyperboloid, with one or two sheets, but their use remains marginal.

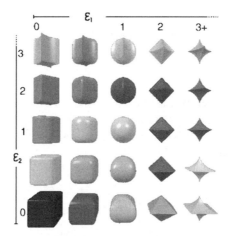

Figure 1.12 Superellipsoid examples and associated ε_1 and ε_2 values. Data Courtesy of L Chevalier

Superquadrics, and more particularly superellipsoids, have a high representational power relative to their small number of parameters; the model is extremely compact. The fact of having both implicit and parametric formulations is also quite interesting, and enables the model to be adapted to the application. Indeed, intersection or blending algorithms are easy to handle using the implicit equation while the parametric formula facilitates sampling or meshing. These surfaces are quite popular in computer graphics (Jaklic and Solina 2003; Jaklic *et al.* 2000; Montiel *et al.* 1998).

In spite of a rather high representational power considering the extremely small number of parameters, it still remains much lower than those of surface-based models (parametric, implicit, subdivision). Indeed it is quite difficult to perfectly model an arbitrary object even by considering a whole graph of superquadrics. Thus, many authors have introduced deformation techniques in order to increase the flexibility of this model, and particularly to reduce its heavy symmetry constraints.

Barr (1984) proposes a set of global deformations: torsion, folding, etc. Terzopoulos and Metaxas (1991) add local deformation coefficients. Bardinet *et al.* (1994) consider control-point-based deformations and Zhou and Kambhamettu (2001) replace the exponents ε_1 and ε_2 by functions depending on latitudinal and longitudinal angles. Blanc and Schlick (1996) introduce a new model, ratioquadrics, in which the exponential functions are replaced by numerically simpler ones. Lastly, DeCarlo and Metaxas (1998) propose blending operations between superquadrics.

Supershapes

Supershapes were recently introduced by Gielis *et al.* (2003). Supershapes can be considered as a generalization of superquadrics, allowing the modeling of asymmetrical shapes. Like superquadrics, they have both parametric and implicit formulations.

Parametric formulation:

$$S(\eta, \mu) = \begin{bmatrix} r_1(\mu)r_2(\eta)\cos(\eta)\cos(\mu) \\ r_1(\mu)r_2(\eta)\cos(\mu)\sin(\mu) \\ r_2(\eta)\sin(\eta) \end{bmatrix} ; -\frac{\pi}{2} \leq \eta \leq \frac{\pi}{2}, -\pi \leq \mu \leq \pi \tag{1.14}$$

Implicit formulation:

$$F(x, y, z) = 1 - \frac{x^2 + y^2 + r_1^2(\mu)z^2}{r_1^2(\mu)z^2 r_2^2(\eta)z^2} \tag{1.15}$$

with

$$r(\mu) = \frac{1}{\sqrt[n_1]{\frac{1}{a}\cos\left(\frac{m\mu}{4}\right)^{n_2} + \frac{1}{b}\sin\left(\frac{m\mu}{4}\right)^{n_3}}} \tag{1.16}$$

n_1, n_2 and n_3 are three shape coefficients and the term $m\mu/4$ allows a rational or irrational number of symmetry, while a and b control the size of the shape. Figure 1.13 illustrates some examples of supershapes.

Figure 1.13 Supershape examples. (From the work of Fougerolle *et al.* (2006)). Reproduced with kind permission of Springer Science and Business Media

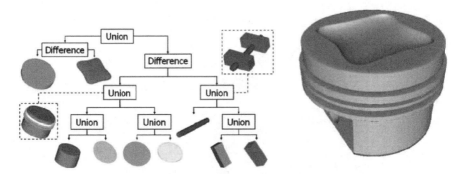

Figure 1.14 Combination of several deformed supershapes (left) and corresponding constructed mesh (right) by the algorithm of Fougerolle *et al.* (2005). Reproduced by permission of IEEE

This model is very recent and exhibits a higher representational power than superquadrics (by breaking the symmetry constraints) while keeping the same benefits of both implicit and parametric formulations. Several authors have investigated the use of combinations of deformed supershapes to represent more and more complex 3D models. Figure 1.14 illustrates a mechanical model composed with 9 supershapes associated with boolean operations. Fougerolle *et al.* (2005) introduced a recent algorithm to accurately polygonize such supershape combinations, and proposed a new formulation for the implicit representation (Fougerolle *et al.* 2006).

Hyperquadrics

Hyperquadrics (Vaerman *et al.* 1997) are a generalization of superquadrics, they enable the representation of a higher variety of shapes and do not have symmetry constraints. However, they are more complex to handle and, contrary to superquadrics, they only have an implicit formulation (Fougerolle *et al.* 2006):

$$\sum_i S_i + e^{\sum_j S_j} = 0 \qquad (1.17)$$

where S is a superquadric equation.

Although hyperquadrics have a higher representational power than superquadrics, their use remains marginal. Their high complexity and the lack of parametric formulations make them of limited interest. Like superquadrics,

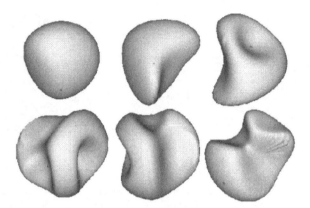

Figure 1.15 Hybrid hyperquadrics, introduced by Cohen and Cohen (1996). Reprinted from Cohen and Cohen, 1996, with permission from Elsevier

these primitives are used especially for organic part reconstruction and simulation in the medical field. The difficulty in controlling the shape has led researchers to introduce local deformations; thus, Cohen and Cohen (1996) introduced the hybrid hyperquadrics, by adding an exponential term to the equation. Examples of their model are illustrated in Figure 1.15.

Constructive solid geometry (CSG)

Description

This model represents a 3D object by a set of linear transformations and boolean operations (union, intersection, difference) applied on elementary primitives (sphere, torus, cone, cubes). Thus the object is represented by a tree (called a *constructive tree*), with a primitive associated with each leaf and an operator associated with each non-terminal node. Figure 1.16 illustrates such a model.

Evaluation

Constructive solid geometry is only used in CAD to represent mechanical parts. The model is extremely compact and permits representation of the whole modeling process. However, this model is very hard to display, and involves discontinuity problems at the joints between primitives. Moreover, there is no associated standard format, and from one piece of CAD software to

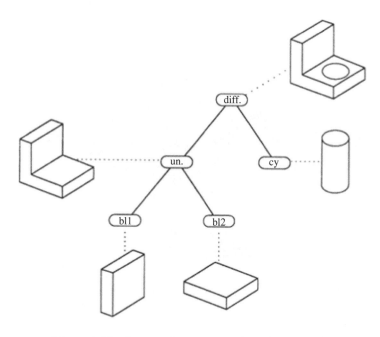

Figure 1.16 Constructive representation of a 3D object

another, operators and primitives are not the same. Further, the geometrical and physical properties are quite complex to calculate. Lastly, the inverse problem of constructing a CSG tree from a set of 3D data (3D point cloud, or 3D mesh) is almost impossible to handle.

Discrete model

Description

This model decomposes the 3D space into volume elements (called voxels). Thus a 3D object is represented by the set of voxels constituting (or intersecting) its volume. An example of a discrete 3D object is illustrated in Figure 1.17.

Evaluation

This discrete model is quite simple to construct and facilitates boolean operations and calculation of physical properties. Moreover, several authors have

Figure 1.17   Discrete representation of the 3D object *Al*. Courtesy of I Sivignon

generalized differential operators to such models (Coeurjolly and Klette 2004; Flin *et al.* 2005). This representation is used particularly in the medical field, since it comes directly from tomographic reconstruction.

However, this model is very expensive in terms of memory and gives only an approximation of the 3D object (arising from its discrete nature) according to a certain level of precision. Moreover, algorithms of ray tracing, meshing or surface or planes extraction are quite complex problems (Kobbelt *et al.* 2001; Sivignon *et al.* 2004). Figure 1.17 illustrates digital planes retrieved using the algorithm from Sivignon *et al.* (2004) (where a color is used to represent a plane).

1.2   3D Data Source

There are several ways to produce 3D content, but basically we can distinguish two principal classes:

- The *acquisition*, which reproduces a real-world object, like a camera for 2D images.
- The *manual creation*, using specific software, similar to a painting of 2D images.

1.2.1   Acquisition

The main objective is to create a 3D object which reproduces a given real-world object, like a photograph represents a real-world image. There are mainly two classes of technologies: those which analyse and reconstruct the *surface* of the object (e.g. 3D scanner) and those which reconstruct its *volume* (e.g. tomography).

3D scanner

A 3D scanner works like a camera, but instead of collecting color information about a surface, it collects *distance* information. Basically, each scan produces a picture where each pixel contains the distance to the surface; this picture is called a depth (or *range*) image. Usually to analyze properly a complex 3D object, multiple scans (i.e. range images) from different directions are necessary. These *range images* are then merged together to produce a cloud of points representing the surface (see Figure 1.1(a)). This operation is called *registration* or *alignment* (Salvi *et al*. 2007) and consists of bringing the range images, coming from different directions, within the same 3D coordinate system. Figure 1.18 illustrates examples of registration.

There are several types of 3D scanners:

- *Laser scanners* emit a beam toward the object and detect its reflection. There are basically two technologies. The *time-of-flight* (or *laser range-finder*) scanner measures the time taken by the beam to be reflected by the target and returned to the sender and then deduces the distance according to the speed of light. This technology is not quite precise but can operate

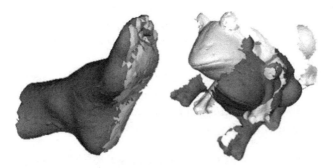

Figure 1.18   Examples of registration of two range images: correct (left) and incorrect (right). Reproduced from Salvi *et al*. 2007.

Figure 1.19   Triangular mesh of the north of Corsica, reconstructed according to a digital
elevation model. Courtesy of DISI by the AIM@SHAPE Shape repository

over very long distances and thus is particularly suitable for measuring
large scenes. For example, with this technique, the *NASA Shuttle Radar
Topographic Mission* provides digital elevation models for over 80 % of
the globe; Figure 1.19 illustrates a triangular mesh of the north of Corsica,
reconstructed according to a digital elevation model. The other technology
is the *triangulation* scanner which emits a beam toward the object and
uses a camera to analyze the location of the laser dot on the object. The
laser dot, the camera and the laser emitter form a triangle which locates
very precisely the corresponding 3D point (accuracy of several microme-
ters). However, such a technique is limited to a depth of several meters.
Figure 1.20 illustrates the laser scanning of the head of Michelangelo's
David and the corresponding 3D model, from the Digital Michelangelo
Project (Levoy *et al.* 2000).

- *Contact scanners* (or *coordinate measuring machines*) use a small arm
(automatic or manual) to touch the 3D object and record the corresponding
3D points. However, such a technique can damage the object (because of
the contact) and is much slower than laser scanners.

There are also some systems that are able to digitize a 3D model without laser
scanning or contact, but only with several cameras using stereo reconstruction
techniques. Hence two images of the same object, associated with adapted

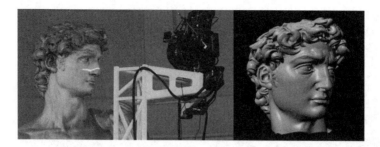

Figure 1.20 Scanning of the head of Michelangelo's David with a triangulation laser scanner (left) and corresponding 3D model (right). Reproduced from the Digital Michelangelo Project (Levoy *et al*. 2000), Stanford University, Copyright © Marc Levoy – Paul Debevec

lighting conditions, are able to reconstruct the 3D geometry, under some shape constraints.

Scanning technology is particularly used for:

- Cultural heritage (see Figure 1.20), to preserve in digital form historical pieces or sites.
- Reverse engineering, which basically consists of scanning a mechanical part for instance and then reconstructing the corresponding CAD model (typically a set of parametric surfaces).
- Geographic information science, where the objective is to model 3D geographical information, like terrain models (see Figure 1.19), for applications in cartography or spatial planning.

Tomography

Tomography is a technique that consists of reconstructing the volume of a 3D object from a set of external measures which often involve emitting a certain signal through the object toward a sensor and analyzing the response. The nature of the retrieved volumetric data depends on the nature of the signal and of the sensors. For instance, the *computed axial tomography scanner* (or CT scan) measures the ability of the material to absorb X-rays; this scanner sends X-ray beams through the object, and a sensor measures their residual intensity. Similarly *echography* measures the acoustic impedance of a material by measuring the attenuation of ultrasound sent through the object. Another example is the *nuclear magnetic resonance imaging* (MRI) scanner which measures the

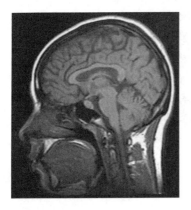 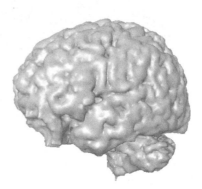

Figure 1.21 A 3D magnetic resonance image of the head, vertical cross-section, and the result after segmentation. The brain 3D model is Courtesy of INRIA by the AIM@SHAPE Shape repository

proton density and structure of a material. The object is placed in a powerful magnetic field. The hydrogen nuclei are excited and then relax by emitting an electromagnetic pulse which is measured by a sensor. Lastly, *positron emission tomography* (PET) works by measuring the radioactive emission of a tracer injected in the object (or the body) for analysis.

The tomography technology produces an image of the considered physical measures (X-ray absorption, ultra sound response, etc.) which has to be converted to 3D data by a mathematical *reconstruction* process. Unlike 3D laser scanners, the tomography technology does not provide a cloud of points but a volumetric model in the form of a 3D grid where each element contains a kind of density value or other information. To be properly utilized, these data often have to be segmented to obtain a 3D binary grid or a polygonal or tetrahedral mesh representing a specific part. Figure 1.21 illustrates a section of a 3D MRI image and the segmentation result.

Tomography technologies mostly concern medical imagery, where analyzing the interior of an organ is critical to detect eventual diseases or pilot surgical operations. Another application of tomography is in geophysics to analyze some aspects of the Earth.

1.2.2 Manual Creation

A 3D content can be obtained by reproducing a real-world object (*acquisition*, see previous section), but can also be created manually by designers using

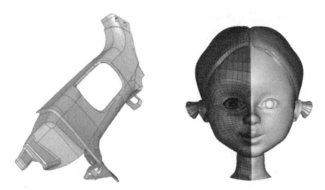

Figure 1.22   Examples of hand-designed 3D objects. Left: CAD object from a Laguna car. Right: Little girl from a 3D modeler (Software Amira)

specific software. Their are two principal types of software: *CAD software* and 3D *modelers*.

CAD software

CAD software helps engineers or architects to design industrial objects, machinery, manufactured content, buildings, etc. This software is based mostly on parametric surfaces and can produce 3D content to a very high precision since this content is then used for manufacturing the real object. An example of a CAD model composed with parametric surfaces is given in Figure 1.22 (left); the boundaries between parametric surfaces are indicated by the solid lines.

3D modelers

Whereas CAD software aims at designing high-precision technical objects, 3D modelers are rather adapted to produce artistic 3D content and 3D animations. The main applications are videogames and 3D movies. Most 3D modelers are based on a polygonal mesh representation since this model is highly adapted for rendering or simple operations like collision detection etc. Texture and color information are often added to the model (see Figure 1.22, right).

1.2.3 3D Object Databases

A critical issue in the design of new algorithms and evaluation of their performance is the presence of open 3D object databases. However, only a few 3D databases exist presently. Recently, the AIM@SHAPE Shape Repository was created within the Network of Excellence project AIM@SHAPE (Advanced and Innovative Models And Tools for the development of Semantic-based systems for Handling, Acquiring, and Processing knowledge Embedded in multidimensional digital objects, 2004–2007). This repository contains a collection of digitally scanned 3D shapes (3D meshes, range images, volumetric models) and offers some interesting functionalities like multiresolution formats. Another example is the Princeton Shape Benchmark which was created specifically to compare 3D indexing methods, hence the objects are classified into several classes. Lastly, some other databases exist for specific 3D face applications: namely, the $3D_R MA$ database at the Signal and Image Center (SIC) at the Royal Military Academy of Belgium, and the database of the University of Notre Dame (UND) in the United States.

1.3   3D Quality Concepts

Compression, or even watermarking, can introduce degradations of the visual quality of a 3D object. Since the ultimate application of these processes is often to be seen by human beings, the need for efficient tools to measure the loss of quality or the visual differences between 3D objects is critical.

Some geometric measures exist between 3D objects, such as the *Hausdorff* distance, where $e(p, M)$ represents the distance from a point p of the 3D space to a 3D object M:

$$e(p, M) = \min_{\{p' \in M\}} d(p, p') \tag{1.18}$$

with d the Euclidean distance between points p and p'. Then the asymmetric Hausdorff distance between two 3D objects M and M' is:

$$H_a(M, M') = \max_{\{p \in M\}} e(p, M') \tag{1.19}$$

The symmetric Hausdorff distance is then defined as follows:

$$H_s(M, M') = \max \left\{ H_a(M, M'), H_a(M', M) \right\} \tag{1.20}$$

The Hausdorff distance is based on an L_∞ norm. Other distances exist based on L_1 or L_2 norms; for example, the mean distance (L_1) is defined as follows:

$$D_{L1}(M, M') = \frac{1}{\text{Area}(M)} \int_M e(p, M') dM \tag{1.21}$$

These geometric distances are available in many software packages (Aspert *et al.* 2002; Cignoni *et al.* 1998), especially for polygonal meshes, but are not well matched with visual human perception. This *perceptual gap* was investigated a great deal in the field of image processing.

1.3.1 Existing 2D Perceptual Metrics

A review of existing 2D perceptual metrics can be found in Eckert and Bradley (1998). Basically there are two different approaches: *computational* and *ad hoc.*

Computational metrics, like the visible difference predictor (VDP) of Daly (1993), consist of complex numerical models taking into account psychophysical and physiological evidence. These models often rely on the same perceptual attributes (Eckert and Bradley 1998):

- The contrast sensitivity function (CSF) which defines the contrast at which frequency components become just visible.
- The channel decomposition, claiming that human vision consists of several channels selective to spatial frequency and to orientation.
- The masking effect which defines the fact that a signal can be masked by the presence of another signal with similar frequency or orientation.

These attributes lead to filtering operations (according to the CSF), filter bank decomposition, error normalization (masking effect) and error summation across frequency bands and space (usually using the Minkowski metric).

Ad hoc metrics consider simpler mathematical measures, intuitively relating to visual perception and/or introducing penalties for specific artifacts. A typical example is the work of Marziliano *et al.* (2004) which aims at detecting and quantifying blocking and ringing artifacts of JPEG compression.

Recently, Wang *et al.* (2004) introduced an alternative framework for image quality assessment which does not rely on a summation of kinds of perceptual errors, but on the degradation of the structural information.

1.3.2 Existing 3D Perceptual Metrics

Transposing complex *computational* metrics from 2D images to 3D objects is quite difficult, and to the best of our knowledge was only investigated by Ferwerda *et al.* (1997). They proposed a masking model, extending the

Daly VDP, which demonstrates how surface texture can mask the polygonal tessellation.

Most of the work on 3D perceptual distances has focused on three specific applications: realistic rendering, mesh simplification and evaluation of specific processes (compression or watermarking).

The objective of perceptually driven rendering is to determine, according to the location of the observer, which level of detail (LoD) to use to satisfy frame rate and image quality requirements. In fact these methods are based on 2D image perceptual models described in the previous section. Reddy (1996, 2001) analyzed the frequency content in several pre-rendered images to determine the best LoD. In a different way, Bolin and Meyer (1998) and more recently Farrugia and Peroche (2004) used perceptual models to optimize the sampling for ray tracing algorithms. Most of these works concern off line rendering. Luebke and Erikson (1997), Luebke and Hallen (2001) and Dumont *et al.* (2003) presented view-dependent simplification algorithms for real-time rendering, based on the worst cases of imperceptible contrast and spatial frequency changes. Williams *et al.* (2003) extended the work of Luebke *et al.* to textured objects. Although the existing work on realistic rendering is based mostly on 2D image metrics, several authors have considered some kinds of 3D metrics, namely Tian and AlRegib (2004) and Pan *et al.* (2005). Their metrics rely respectively on geometry and texture deviations (Tian and AlRegib 2004) and on texture and mesh resolutions (Pan *et al.* 2005).

Some real 3D metrics (*ad hoc*) are used to control mesh simplification algorithms, which consist of reducing the number of vertices while preserving the visual appearance. Kim *et al.* (2002) stated that human vision is sensitive to curvature changes and proposed a *discrete differential error metric* (DDEM), which is basically the following, between two vertices v and v':

$$\text{DDEM}(v, v') = Q(v, v') + T(v, v') + C(v, v') \qquad (1.22)$$

with Q a quadratic distance, T a normal vector difference and C a discrete curvature difference. In a different way, Howlett *et al.* (2005) pilot their simplification algorithm so as to emphasize visually salient features, determined by an eye tracking system. Lee *et al.* (2005) followed a similar approach but automatically extract the saliency from the input mesh by computing kinds of multiresolution curvature maps.

Recently several authors have investigated the use of perceptual metrics (*ad hoc*) for the evaluation of specific applications. Karni and Gotsman

(2000), in order to evaluate properly their compression algorithm, introduced the *geometric Laplacian* (GL), which measures the smoothness of a vertex v:

$$\text{GL}(v) = v - \frac{\sum_{i \in n(v)} l_i^{-1} v_i}{\sum_{i \in n(v)} l_i^{-1}} \tag{1.23}$$

where $n(v)$ is the set of indices of the neighbors of v, and l_i the Euclidean distance from v to v_i. $\text{GL}(v)$ represents the difference vector between v and its new position after a Laplacian smoothing step. Thus it represents a measure of smoothness: the lower it is, the smoother is the surface around v. Using this Geometric Laplacian, Karni and Gotsman introduced a perceptual metric between two meshes X and Y, with the same connectivity, containing n vertices:

$$\text{GLD}(X, Y) = \frac{1}{2n} \left(\sum_{i=0}^{n-1} \left\| v_i^x - v_i^y \right\| + \sum_{i=0}^{n-1} \left\| \text{GL}(v_i^x) - \text{GL}(v_i^y) \right\| \right) \tag{1.24}$$

where v_i^x and v_i^y are the respective ith vertices from X and Y. In the same way, Drelie Gelasca *et al.* (2005) proposed a new metric based on global *roughness* variation, to measure the perceptual quality of a watermarked mesh. They define the *roughness* as the variance of the difference between a 3D model and its smoothed version, similar to the geometric Laplacian of Karni and Gotsman (2000). Corsini *et al.* (2005) presented a similar *roughness*-based measure. Rondao-Alface *et al.* (2005) presented two other metrics to benchmark watermarking schemes, one based on a measure of distortion between several 2D views, and the other based on the distortion of energy calculated using 2D parameterization of the meshes. Finally Lavoué *et al.* (2006) introduced a structural distortion measure which reflects the visual similarity between two meshes in a general-purpose context: the MSDM (3D Mesh Structural Distortion Measure).

1.3.3 A Perceptual Measure Example

The MSDM, from Lavoué *et al.* (2006), follows the concept of structural similarity recently introduced for 2D image quality assessment by Wang *et al.* (2004). This measure, which aims at reflecting the perceptual similarity between two 3D objects, is based on curvature analysis (mean, standard deviation, covariance) on local windows of the meshes. The authors

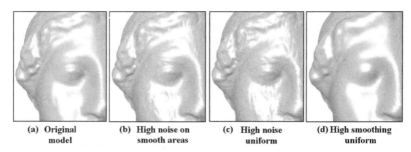

<table>
<tr><td>(a) Original
model</td><td>(b) High noise on
smooth areas</td><td>(c) High noise
uniform</td><td>(d) High smoothing
uniform</td></tr>
</table>

Figure 1.23 (a) Zoom on the original *Venus* object, (b) high noise (maximum deviation = 0.012) on smooth regions (MOS = 8.8, MSDM = 0.64, D_{L1} = 0.16), (c) high noise on the whole object (MOS = 9.4, MSDM = 0.70, D_{L1} = 0.26), (d) high smoothing (30 iterations) on the whole object (MOS = 8.1, MSDM = 0.58, D_{L1} = 0.25)

have evaluated their metric through a subjective experiment: a test corpus was created, containing distorted versions (issuing from noise addition and smoothing operations) of several 3D objects. Examples of distortions are provided in Figure 1.23 for the *Venus* 3D model. Then a pool of human observers were asked to rate the visual similarity of these versions with the corresponding original objects. Thus each object of this corpus is associated with a mean opinion score (MOS) reflecting its subjective similarity, from 0 (identical to the original) to 10 (very different), to the original model and an MSDM distance value ($\in$ [0, 1]). A cumulative Gaussian psychometric curve is then used to provide a nonlinear mapping between the objective and subjective scores.

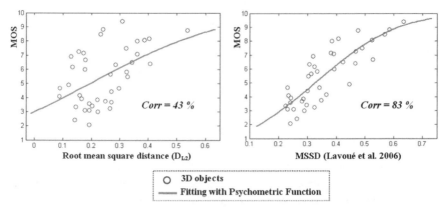

Figure 1.24 Subjective MOS vs. MSDM metric and the root mean square distance (D_{L2}) for 42 test objects

Figure 1.23 provides a first intuitive example: the *noised* object in (b) is associated with a higher MOS than the *smoothed* object in (d) (8.8 vs. 8.1). That seems intuitively normal since the smoothed model appears visually less distorted than the noised one. The MSDM reflects well this subjective opinion since, for (b), it is also higher than for (d) (0.64 vs. 0.58). On the contrary, the geometric mean distance (see Equation (1.21)) does not reflect this subjective opinion at all (0.16 vs. 0.25).

Figure 1.24 presents results of the psychometric curve fitting (cumulative Gaussian) between subjective and objective scores (root mean square distance (L_2) and MSDM). The correlation (Spearman) is better for MSDM (83 % vs. 43 %) which predicts well the subjective scores (MOS).

1.4   Summary

Technological advances in the fields of telecommunications, computer graphics and multimedia during the last decade have contributed to the evolution of digital data being manipulated, visualized and transmitted over the Internet. Nowadays, 3D data constitute the new emerging multimedia content.

However, 3D objects are more complex to handle than other multimedia data, such as audio signals or 2D images, since there are many different representations for such objects.

Three representation schemes are mainly used:

- **Polygonal mesh**: with this model, the object is represented by its boundary surface which is composed of a set of planar faces. Basically a polygonal mesh contains a set of 3D points (*vertices*) which are linked by *edges* to form a set of polygonal *facets*.
- **Parametric surface**: this family of 3D surfaces (including Bézier, B-spline and NURBS surfaces) is used particularly for computer-aided design (CAD). This model enables us to define mathematically exact surfaces, contrary to polygonal meshes which only represent an approximation.
- **Discrete representation**: this model does not represent the 3D object in Euclidean space, but in a 3D grid similar to the 2D image representation. Each element of the grid is a voxel (for volumetric pixel). Thus the object is represented by the set of voxels constituting its volume.

Several other representation models exist (implicit surface, subdivision surface, superquadric, CSG, etc.), but their utilization concerns more specific domains.

This 3D content comes from two main sources:

- The *acquisition*, using a 3D scanner or tomography for instance, which reproduces a real-world object, like a camera does for a 2D image.
- The *manual creation*, using specific software, similar to a painting of a 2D image.

The 3D models are subject to a wide variety of processing operations such as compression, simplification, watermarking or indexing, which may introduce some geometric artifacts on the shape. The main issue is to maximize the compression/simplification ratio or the watermark strength while minimizing these visual degradations.

In this context the need for efficient tools to measure the loss of quality or the visual difference between 3D objects becomes critical. However, classical metrics based on geometric differences like the Hausdorff distance do not match well with human visual perception. This perceptual gap has just started to be investigated for 3D models.

References

Akkouche S and Galin E 2001 Adaptive implicit surface polygonization using marching triangles. *Computer Graphics Forum* **20**(2), 67–80.

Alexa M, Gross M, Pauly M, Pfister H, Stamminger M and Zwicker M 2004 Point-based computer graphics. *ACM SIGGRAPH course notes*, p. 7.

Angelidis A and Cani MP 2002 Adaptive implicit modeling using subdivision curves and surfaces as skeletons. *Solid Modelling and Applications*, pp. 45–52.

Aspert N, Santa-Cruz D and Ebrahimi T 2002 Mesh: measuring error between surfaces using the Hausdorff distance. *IEEE International Conference on Multimedia and Expo (ICME)*, pp. 705–708.

Barbier A, Galin E and Akkouche S 2005 A framework for modeling, animating, and morphing textured implicit models. *Journal of Graphical Models* **67**(3), 166–188.

Bardinet E, Cohen LD and Ayache N 1994 Fitting 3D data using superquadrics and free-form deformations. *International Conference on Pattern Recognition*, pp. 79–83.

Barr AH 1981 Superquadrics and angle preserving transformations. *IEEE Computer Graphics and Applications* **1**(1), 11–23.

Barr AH 1984 Global and local deformations of solid primitives. *Computer Graphics* **18**(3), 21–30.

Bézier P 1972 *Numerical control: Mathematics and applications.* John Wiley & Sons' Ltd.

Biederman I 1985 Human image understanding: recent research and a theory. *Computer Vision, Graphics, and Image Processing* **32**, 29–73.

Binford T 1971 Visual perception by computer. *IEEE Conference on Systems and Control.*

Blanc C and Schlick C 1996 Ratioquadrics: an alternative model for superquadrics. *The Visual Computer* **12**(8), 420–428.

Blinn JF 1982 A generalization of algebraic surface drawing. *ACM Transactions on Graphics* pp. 235–256.

Bloomenthal J and Shoemake K 1991 Convolution surfaces. *Computer Graphics* **25**(4), 251–256.
Bloomenthal J and Wyvill B 1990 Interactive techniques for implicit modeling. *Computer Graphics* **24**(2), 109–116.
Bolin M and Meyer G 1998 A perceptually based adaptive sampling algorithm. *ACM SIGGRAPH*, pp. 299–309.
Catmull E and Clark J 1978 Recursively generated B-spline surfaces on arbitrary topological meshes. *Computer-Aided Design* **10**(6), 350–355.
Cignoni P, Occhini C and Scorpigno R 1998 Metro: measuring error on simplified surfaces. *Computer Graphics Forum* **17**(2), 167–174.
Coeurjolly D and Klette R 2004 A comparative evaluation of length estimators of digital curves. *IEEE Transactions on Pattern Analysis and Machine Intelligence* **26**(2), 252–258.
Cohen I and Cohen LD 1996 A hybrid hyperquadric model for 2D and 3D data fitting. *Computer Vision and Image Understanding* **63**(3), 527–541.
Corsini M, Drelie Gelasca E and Ebrahimi T 2005 A multi-scale roughness metric for 3D watermarking quality assessment. *Workshop on Image Analysis for Multimedia Interactive Services.*
Daly S 1993 The visible differences predictor: an algorithm for the assessment of image fidelity. In *Digital Images and Human Vision*, A. B. Watson, Ed. MIT Press, Cambridge, MA, 179–206.
DeCarlo D and Metaxas D 1998 Shape evolution with structural and topological changes using blending. *IEEE Transactions on Pattern Analysis and Machine Intelligence* **20**, 1186–1205.
Doo D and Sabin M 1978 Behavior of recursive division surfaces near extraordinary points. *Computer Aided Design* **10**, 356–360.
Drelie Gelasca E, Corsini M and Ebrahimi T 2005 Objective evaluation of the perceptual quality of 3D watermarking. *IEEE International Conference on Image Processing*, pp. 241–244.
Dumont R, Pellacini F and Ferwerda J 2003 Perceptually-driven decision theory for interactive realistic rendering. *ACM Transactions on Graphics* **22**(2), 152–181.
Eckert M and Bradley A 1998 Perceptual quality metrics applied to still image compression. *Signal Processing* **70**(3), 177–200.
Farin G 1993 *Curves and surfaces for CAGD: A practical guide*. Academic Press.
Farrugia J and Peroche B 2004 A progressive rendering algorithm using an adaptive perceptually based image metric. *Computer Graphics Forum* **23**(3), 605–614.
Ferwerda J, Pattanaik S, Shirley P and Greenberg D 1997 A model of visual masking for computer graphics. *ACM SIGGRAPH*, pp. 143–152.
Flin F, Brzoska JB, Lesaffre B, Coléou C, Lamboley P, Coeurjolly D, Teytaud O, Vignoles G and Delesse JF 2005 An adaptive filtering method to evaluate normal vectors and surface areas of 3D objects. Application to snow images from X-ray tomography. *IEEE Transactions on Image Processing* **14**(5), 585–596.
Fougerolle Y, Gribok A, Foufou S, Truchetet F and Abidi M 2005 Boolean operations with implicit and parametric representation of primitives using R-functions. *IEEE Transactions on Visualization and Computer Graphics* **11**(5), 529–539.
Fougerolle Y, Gribok A, Foufou S, Truchetet F and Abidi M 2006 Radial supershapes for solid modeling. *Journal of Computer Science and Technology* **21**(2), 238–243.
Gielis J, Beirinckx B and Bastiaens E 2003 Superquadrics with rational and irrational symmetry. *ACM Symposium on Solid Modeling and Applications*, pp. 262–265.
Han S, Goldof D and Bowyer K 1993 Using hyperquadrics for shape recovery from range data. *IEEE International Conference on Computer Vision*, pp. 492–496.
Hoppe H, DeRose T, Duchamp T, Halstead M, Jin H, McDonald J, Schweitzer JE and Stuetzle W 1994 Piecewise smooth surface reconstruction. *ACM SIGGRAPH*, pp. 295–302.
Hornus S, Angelidis A and Cani MP 2003 Implicit modelling using subdivision-curves. *The Visual Computer* **19**(2–3), 94–104.
Howlett S, Hamill J and O'Sullivan C 2005 Predicting and evaluating saliency for simplified polygonal models. *ACM Transactions on Applied Perception* **2**(3), 286–308.

Jaklic A and Solina F 2003 Moments of superellipsoids and their application to range image registration. *IEEE Transactions on Systems, Man, and Cybernetics* **33**(4), 648–657.

Jaklic A, Leonardis A and Solina F 2000 *Segmentation and Recovery of Superquadrics*. Kluwer Academic.

Karni Z and Gotsman C 2000 Spectral compression of mesh geometry. *ACM SIGGRAPH*, pp. 279–286.

Keren D, Cooper DB and Subrahmonia J 1994 Describing complicated objects by implicit polynomials. *IEEE Transactions on Pattern Analysis and Machine Intelligence* **16**(1), 38–53.

Kim S, Kim S and Kim C 2002 Discrete differential error metric for surface simplification. *Pacific Graphics*, pp. 276–283.

Kobbelt L 1996 Interpolatory subdivision on open quadrilateral nets with arbitrary topology. *Computer Graphics Forum* **15**(3), 409–420.

Kobbelt L and Botsch M 2004 A survey of point-based techniques in computer graphics. *Computers and Graphics* **28**(6), 801–814.

Kobbelt LP, Botsch M, Schwanecke U and Seidel HP 2001 Feature-sensitive surface extraction from volume data. *ACM SIGGRAPH*, pp. 27–66.

Krishnan S and Manocha D 1997 An efficient surface intersection algorithm based on lower-dimensional formulation. *ACM Transactions on Graphics* **16**(1), 74–106.

Lavoué G, Drelie Gelasca E, Dupont F, Baskurt A and Ebrahimi T 2006 Perceptually driven 3D distance metrics with application to watermarking. *SPIE Applications of Digital Image Processing XXIX*.

Lee C, Varshney A and Jacobs D 2005 Mesh saliency. *ACM SIGGRAPH*, pp. 659–666.

Loop C 1987 Smooth subdivision surfaces based on triangles. Master's thesis, Utah University.

Lorensen WE and Cline HE 1987 Marching cubes: a high resolution 3D surface construction algorithm. *Computer Graphics* **21**(4), 163–168.

Luebke D and Erikson C 1997 View-dependent simplification of arbitrary polygonal environments. *ACM SIGGRAPH*, pp. 199–208.

Luebke D and Hallen B 2001 Perceptually driven simplification for interactive rendering. *Eurographics Workshop on Rendering Techniques*, pp. 223–234.

Marc Levoy, Kari Pulli, Brian Curless, Szymon Rusinkiewicz, David Koller, Lucas Pereira, Matt Ginzton, Sean Anderson, James Davis, Jeremy Ginsberg, Jonathan Shade, and Duane Fulk. The Digital Michelangelo Project: 3D Scanning of Large Statues. Proceedings of ACM SIGGRAPH 2000. pp. 131–144,

Marziliano P, Dufaux F, Winkler S and Ebrahimi T 2004 Perceptual blur and ringing metrics: Application to JPEG2000. *Signal Processing: Image Communication* **19**(2), 163–172.

Montiel E, Aguado AS and Zaluska E 1998 Surface subdivision for generating superquadrics. *The Visual Computer* **14**(1), 1–17.

MPEG-4 2002 ISO-IEC 14496-16. Coding of audio-visual objects: animation framework extension (AFX).

Murakami S and Ichihara H 1986 On a 3D display method by metaball technique. *Transactions of the Institute of Electronics, Information and Communication Engineers of Japan* **J70**(8), 1607–1615.

Muraki S 1991 Volumetric shape description of range data using *Blobby Model*. *Computer Graphics* **25**(4), 227–235.

Nguyen QL and Levine MD 1996 Representing 3D objects in range images using geons. *Computer Vision and Image Understanding* **63**(1), 158–198.

Nishimura H, Hirai M, Kawai T, Kawata T, Shirakawa I and Omura K 1985 Object modeling by distribution function and a method of image generation. *Transactions of the Institute of Electronics and Communication Engineers of Japan* **J68**(4), 718–725.

Pan Y, Cheng I and Basu A 2005 Quality metric for approximating subjective evaluation of 3-D objects. *IEEE Transactions on Multimedia* **7**(2), 269–279.

Peters J and Reif U 1997 The simplest subdivision scheme for smoothing polyhedra. *ACM Transactions on Graphics* **16**(4), 420–431.

Ponce J, Chelberg DM and Mann WB 1989 Invariant properties of straight homogeneous generalized cylinders and their contours. *IEEE Transactions on Pattern Analysis and Machine Intelligence* **11**(9), 951–966.

Reddy M 1996 Scrooge: Perceptually-driven polygon reduction. *Computer Graphics Forum* **15**(4), 191–203.

Reddy M 2001 Perceptually optimized 3D graphics. *IEEE Computer Graphics and Applications* **21**(5), 68–75.

Rondao-Alface P, De Craene M and Macq B 2005 Three-dimensional image quality measurement for the benchmarking of 3D watermarking schemes. *Electronic Imaging, Security, Steganography, and Watermarking of Multimedia Contents*, pp. 230–240.

Salvi J, Matabosch C, Fofi D and Forest J 2007 A review of recent range image registration methods with accuracy evaluation. *Image and Vision Computing*, **25**(5), 578–596.

Schaefer S and Warren J 2005 On C^2 triangle/quad subdivision. *ACM Transactions on Graphics* **24**(1), 28–36.

Sivignon I, Dupont F and Chassery JM 2004 Decomposition of a 3D discrete object surface into discrete plane pieces. *Algorithmica* **38**, 25–63.

Stam J and Loop C 2003 Quad/triangle subdivision. *Computer Graphics Forum* **22**(1), 79–85.

Terzopoulos D and Metaxas D 1991 Dynamic 3D models with local and global deformations: deformable superquadrics. *IEEE Transactions on Pattern Analysis and Machine Intelligence* **13**(7), 703–714.

Tian D and AlRegib G 2004 FQM: a fast quality measure for efficient transmission of textured 3D models. *ACM Multimedia*, pp. 684–691.

Vaerman V, de Sola Fabregas C and Menegaz G 1997 The hyperquadrics: an efficient parametric surface representation. Technical Report LTS 97.10, Signal Processing Lab, Swiss Federal Institute of Technology.

Wang Z, Bovik A, Sheikh H and Simoncelli E 2004 Image quality assessment: from error visibility to structural similarity. *IEEE Transactions on Image Processing* **13**(4), 1–14.

Warren J and Weimer H 2002 *Subdivision methods for geometric design: A constructive approach*. Morgan Kaufmann.

Williams N, Luebke D, Cohen J, Kelley M and Schubert B 2003 Perceptually guided simplification of lit, textured meshes. *ACM Symposium on Interactive 3D Graphics*, pp. 113–121.

Wyvill B, Guy A and Galin E 1999 Extending the CSG tree (warping, blending and boolean operations in an implicit surface modeling system). *Computer Graphics Forum* **18**(2), 149–158.

Wyvill G, McPheeters C and Wyvill B 1986 Data structure for soft objects. *The Visual Computer* **2**, 227–234.

Zeroug M and Nevatia R 1996 Three-dimensional descriptions based on the analysis of the invariant and quasi-invariant properties of some curved-axis generalized cylinders. *IEEE Transactions on Pattern Analysis and Machine Intelligence* **18**(3), 237–253.

Zhou L and Kambhamettu C 2001 Extending superquadrics with exponent functions: modeling and reconstruction. *Journal of Graphical Models* **63**(1), 1–20.

Zorin D, Schroder P and Sweldens W 1996 Interpolating subdivision for meshes with arbitrary topology. *ACM SIGGRAPH*, pp. 189–192.

2

3D Compression

Guillaume Lavoué, Florent Dupont, Atilla Baskurt

2.1 Introduction

Technological advances in the fields of telecommunications, computer graphics and multimedia during the last decade have contributed to the evolution of digital data being manipulated, visualized and transmitted over the Internet. Nowadays, 3D data constitute the emerging multimedia content. In this context the need for efficient tools to reduce the size of this 3D content, mostly represented by polygonal meshes, becomes even more acute, particularly to reduce the transmission time for low-bandwidth applications. That is precisely the objective of compression techniques.

A compression algorithm often consists of two main parts:

- Finding a concise/synthetic representation of the 3D model, even if it implies some loss of precision or some errors regarding the original object.
- Suppressing the redundancy using classical coding techniques, such as quantization, prediction and entropic encoding.

3D Object Processing: Compression, Indexing and Watermarking   J.-L. Dugelay, A. Baskurt & M. Daoudi
© 2008 John Wiley & Sons, Ltd

2.2 Basic Review of 2D Compression

In this section we recall some basic compression notions, using the special case of 2D images. A wide community of researchers has been working on compression methods since the 1980s. Important advances allow the transmission of digital television today over a reduced bandwidth ADSL in homes. The well-known JPEG standard for the compression of still images is largely the result of efforts of this part of the scientific community since 1987.

2.2.1 Outline of a Generic Compression Scheme

The main goal of a compression method is to find an optimal compromise between the bit rate and the distorsion: how do we obtain a minimal bit rate while preserving high visual quality (minimization of the whole distortion)? Some methods aim at a perfect reconstruction of the original data with limited compression ratios (lossless compression). When an application requires low bit rates, a controlled loss of information is necessary (lossy compression) in order to enhance these bit rates. The losses can take the form of blocking effects, reduced color quality, or slight local deformations or imprecision on the reconstruction of a 3D object.

A generic compression scheme is composed of three blocks: redundancy detection and reduction, quantization, and coding (see Figure 2.1).

Redundancy detection and reduction

Redundancy can be defined as the amount of similarity between the symbols generated by an information source. It may be spatial (in neighboring pixels or vertices), spectral (between the red, green and blue components, for instance), or temporal (between successive planes of video or animated 3D objects). Compression methods detect the presence of these different types of redundancy using innovative tools (predictions, transforms like wavelets). These tools lead to a compact and decorrelated representation of the information.

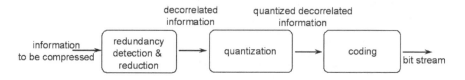

Figure 2.1 Functional scheme of a compression method

Quantization of the decorrelated information

Decorrelated information may be represented by integer, real or complex values in a given dynamic range. These formats and ranges are often incompatible with the average number of bits per symbol associated to the channel capacity. In such cases, quantization methods have to be used.

Two classes of quantization can be distinguished:

- Scalar quantization, applied to scalars (voxel intensity, vertex coordinates, colors).
- Vector quantization, applied to blocks of neighboring vertices or coefficients.

The quantization is the step which introduces the loss of information into the compression scheme.

Coding of the quantized values

The values obtained from the quantization process are represented by a binary code of fixed length (N bits per symbol). The coding stage allows us to reduce the average number of bits allocated to each quantized value. The coding techniques are based on the information theory presented in the next section for 3D objects. They involve no loss of information.

2.2.2 Compression Ratio, Quality Evaluation

The compression ratio is defined as the total number of bits necessary to represent the original information divided by the total number of bits of the binary file which will be stored. In practice, we use the bit rate to measure the compacting efficiency. The bit rate is expressed in bits per element. The latter is a 'pixel' if we are compressing a still or animated image, or a 'sample' for a signal or a vertex if we are dealing with a 3D object.

As a quantitative measure of the quality of the decoded image, the mean squared error (MSE) between the original image X and the decoded one $\hat{X}$ is given by:

$$\text{MSE} = \frac{1}{NM} \sum_{0}^{N-1} \sum_{0}^{M-1} \left[X(i, j) - \hat{X}(i, j) \right]^2 \qquad (2.1)$$

where N/M is the size of the original image, with the number of pixels varying between 0 and 255.

According to the literature, the quality tends to be expressed in dB with a signal-to-noise relationship (PSNR):

$$\mathrm{PSNR} = 10 \log_{10} \frac{255^2}{\mathrm{MSE}} \qquad (2.2)$$

Note that this evaluation is a mathematical one and is not necessarily associated to the visual quality as perceived by an end-user. Several interesting studies have been done on the estimation of visual quality (Daly 1993; Eckert and Bradley 1998).

2.3 Coding and Scalability Basis

This section provides information about several essential concepts of compression, and coding theory. These bases are necessary to understand well the following technical sections on 3D compression.

Firstly the basis of information theory and entropy calculation are presented, together with their specific use in the context of 3D compression. Secondly this section tackles the concept of quantization, and more specifically geometric quantization.

2.3.1 The Basis of Information Theory and Entropic Coding

The *entropy*

The 3D compression schemes which are introduced in the next section often translate a 3D object into a sequence of *symbols* belonging to a specific *alphabet*. For example, the VRML standard representation of a 3D mesh (see Figure 1.4) encodes each of its faces by the corresponding list of vertex indices. Thus each face, of degree d, is encoded by d symbols S_i with $S_i \in [0, 1, \ldots, n-1]$ with n the number of vertices of the mesh.

If we consider a sequence $S = S_0 S_1 \ldots S_{s-1}$ of s symbols, belonging to the alphabet $A = [A_0, A_1, \ldots, A_{a-1}]$ of size a, each symbol can basically be encoded by $\log_2 a$ bits. Thus the sequence will be encoded by $s \times \log_2 a$ bits. Nevertheless, if several symbols are equal to the same A_i, it seems natural to be able to encode the sequence with less than $s \times \log_2 a$ bits.

As a matter of fact, the optimal number of bits necessary to encode a sequence S of symbols from A depends on its probability distribution $P = [P_0, P_1, \ldots, P_{a-1}]$, with P_i the probability of the symbol A_i appearing in the sequence. The optimal average number of bits necessary to encode a symbol, for a given sequence S associated with the distribution P, is given by (Shannon 1948):

$$H(S) = \sum_{i=0}^{n-1} P_i \log_2 \left(\frac{1}{Pi} \right) \tag{2.3}$$

$H(S)$ corresponds to the *entropy* of the sequence, which reflects the amount of *information* or the amount of *uncertainty* in the distribution. The more uniform is the distribution, the higher is the entropy. Hence the entropy verifies the following relation:

$$0 \le H(S) \le \log_2 a \tag{2.4}$$

Entropic coding

Several *entropic* coding schemes have been introduced which exploit the symbol distribution in order to decrease the amount of bits needed to encode a given sequence, and to reach the optimal value given by Equation (2.3). The most used are *Huffman* (Huffman 1952) and *arithmetic* (Witten *et al.* 1987) coding.

The Huffman scheme assigns to each symbol A_i a codeword C_i of variable length. The length depends on the probability P_i: the idea is to associate very short codewords (1 bit for example) to the most frequent symbols, in order to decrease the number of bits of the whole encoded sequence.

To ensure the correct decoding of the sequence, a given codeword must not be the prefix of another. This is done by constructing a binary tree: leaves correspond to symbols and paths from the root to the different leaves provide the associated codewords.

The main drawback of Huffman coding is its incapacity to encode a symbol with less than 1 bit. For example, if we consider a symbol A_i associated with a probability 0.99, the associated codeword will have a 1-bit length whereas the associated entropy is much lower.

The arithmetic coding helps to overcome this limitation and to encode a symbol with a non-integer number of bits. In this scheme the sequence of symbols is encoded by an interval. The principle is basically the following:

the first symbol is encoded by an interval in [0,1] whose length depends on its probability; then with each new symbol, the interval is refined iteratively. Finally every floating number in the final interval encodes the sequence.

These entropic coding algorithms are all more efficient if the entropy of the sequence is low (i.e. symbols are as less equiprobable as possible). Thus the objective of 3D compression algorithms is to transform the data into highly non-uniform sequences of symbols.

2.3.2 Geometric Quantization Concepts

A 3D object, whatever the representation (mesh, parametric surface, etc.), is mostly defined by real values. For instance, these real values can represent the 3D coordinates of the vertices of a polygonal mesh, the nodes of a B-spline surface or the parameters of the potential functions of an implicit model. In an uncompressed form, these real values are usually associated with a 32-bit floating precision that can distinguish between 2^{32} possible values which, in most cases, are far more than is needed for standard applications. The quantization aims at reducing this number of possible values to a representative (i.e. sufficient) number. On the contrary to entropic coding, which is *lossless*, quantization modifies the data in an irreversible way.

The principle of quantization is to project the data onto a grid structure, and then a floating value is encoded only by the index (integer) of the closest point of the grid. Figure 2.2 illustrates the quantization of the geometric coordinates of a 2D polygonal mesh containing seven vertices. Coordinates

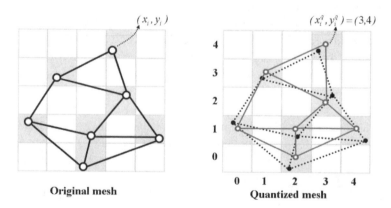

Original mesh   Quantized mesh

Figure 2.2   Geometric uniform quantization of a polygonal mesh on a 5×5 grid

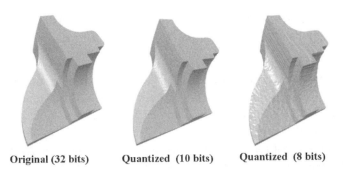

Original (32 bits)   Quantized (10 bits)   Quantized (8 bits)

Figure 2.3   Visual effect of the geometric quantization of a 3D polygonal mesh

(x_i, y_i) of each vertex are quantized on a 5×5 grid, and are thus represented by integers (x_i^q, y_i^q) with x_i^q and $y_i^q \in [0, 4]$. Their quantized coordinates can now be encoded with 3 bits (instead of 32). New positions of the vertices, after decoding, are the centers of the square cells containing them.

This quantization induces a distortion of the original data; the difficulty is to optimize this tradeoff between coding length and distortion. Figure 2.3 shows an example of a 3D polygonal mesh whose vertex coordinates have been quantized with respectively 10 and 8 bits of precision (respectively 1024 and 256 possible values). Whereas the 10-bit quantization is hardly noticeable, the 8-bit one is quite damaging for the appearance of the object.

2.4   Direct 3D Compression

The majority of existing 3D compression algorithms is applied to 3D meshes because this model is by far the most used and is quite large in terms of amount of data. Among the other existing 3D models, only parametric surfaces and voxels have been considered by the 3D compression community. This chapter presents the existing direct compression methods for these three representation models, and particularly for polygonal meshes on which a lot of work has been done in the last 10 years.

2.4.1   Polygonal Mesh Compression

Due to the predominance of polygonal meshes in computer graphics and 3D modeling, and due to the large amount of data that they require, many compression models have been developed in the last decade (the reader can refer to the reviews by Gotsman *et al.* (2002) and Alliez and Gotsman (2005)). They can be separated into three main categories:

- **Simplification**: These algorithms aim to transform the input mesh into a new mesh containing less faces, vertices and edges, but representing the same shape. This is not *compression* in the strict sense of the word, but rather *data reduction*.
- **Single rate compression**: The objective is to suppress the redundancy of the original data. In this class of algorithm, the whole 3D mesh is decoded and reconstructed at the end of the transmission.
- **Progressive compression**: During the decoding, the 3D mesh is reconstructed incrementally, providing access to intermediate states of the object. These progressive (or *multiresolution*) algorithms first transmit a low-resolution version of the mesh, followed by iterative refinement information which allows the construction of finer and finer versions.

Mesh simplification

The mesh simplification algorithms can be classified into three main categories: face merging, incremental decimation and remeshing.

Face merging

This class of algorithms puts the faces together in connected regions. A new mesh is created by replacing each connected region by a plane face. Thus this new mesh possesses a number of faces lower than the original one; however, the faces often have a high degree. The existing algorithms differ in the criteria used to construct the regions. Garland *et al.* (2001) used planarity criteria to incorporate faces in the regions. Sander *et al.* (2001) considered a similar method, while Levy *et al.* (2002) considered curvature information. More recently, Cohen-Steiner *et al.* (2004) minimized a global metric, based on the partitioning algorithm from Lloyd (1982); their results are quasi-optimal for a given number of regions.

Incremental decimation

These incremental algorithms remove vertices iteratively from the mesh. Vertices to be removed are chosen according to several distortion minimization criteria. There exist several topological operations which permit

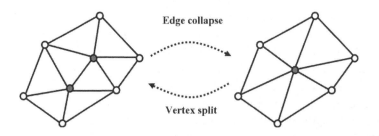

Figure 2.4 Topological operators *edge collapse* and *vertex split*

removal of a vertex while keeping a coherent mesh (i.e. manifold) and while preserving the topology. These operators are called *Euler operators*, the most popular being the *edge collapse* which considers two adjacent vertices and removes the edge connecting them, so the vertices are merged into only one. Figure 2.4 illustrates the *edge collapse* and its dual operator, the *vertex split*. Among the existing decimation approaches, let us cite the reference works of Schroeder *et al.* (1992), Klein *et al.* (1996), Lindstrom and Turk (1998) and Garland and Heckbert (1997) (the most used algorithm). These algorithms differ especially in the distortion measurement which goes from the simple scalar distance (Schroeder *et al.* 1992) or *Hausdorff* distance (Klein *et al.* 1996) to the complex *quadric error metric* (Garland and Heckbert 1997).

Remeshing

Whereas the two preceding classes of approaches are based on the original elements of the input mesh (vertices, faces), the remeshing approaches sample new points on the input mesh and builds the simplified mesh on these new vertices, after reconstructing the topology (by triangulation for instance). The main advantage of these approaches is that the new mesh is independent of the old connectivity. The existing methods differ in the way that they sample the new vertices, i.e. according to the curvature (Turk 1992), the most uniformly possible (Valette and Chassery 2004), or on the contrary by taking into account the anisotropy of the surface (Alliez *et al.* 2003). Once the new vertices have been sampled, the construction of a correct connectivity (set of edges and faces) is not easy (which points can be connected together?). To simplify this step, many approaches carry out this connectivity construction in a 2D global parametric space (Alliez *et al.* 2002, 2003; Lee *et al.* 1998); these methods give very good results but are limited by the necessary parameterization step (spreading out of the mesh in a parametric 2D space) which requires many calculations and often implies strong constraints for the mesh (genus 0, no holes, etc.). Thus, Guskov *et al.* (2000) proposed an approach implying a less complex local parameterization. Some authors have avoided this parameterization step: Kobbelt *et al.* (1999) made an initial base mesh converge toward the input 3D model (*shrink wrapping* approach); Valette and Chassery (2004) assembled the faces in an approximation of a centroidal Voronoi diagram. The centers of the Voronoi cells become the new vertices which are then triangulated. Peyré and Cohen (2005) presented a similar approach: the faces are assembled according to geodesic distance calculations. Finally, several recent approaches are investigating complete quadrangulations of arbitrary meshes (Boier-Martin *et al.* 2004; Dong *et al.* 2006; Ray *et al.* 2006; Tong *et al.* 2006). Figure 2.5 illustrates the algorithm of Alliez *et al.* (2003) which remeshes the original object according to its lines of curvature.

Single rate compression

Most of the existing single rate compression algorithms follow the same scheme. The geometry and the connectivity are encoded separately but not independently; the connectivity is encoded by a region growing approach

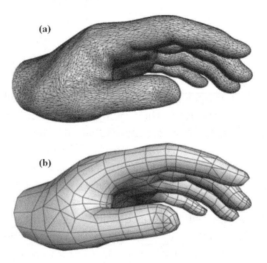

Figure 2.5 Remeshing algorithm: (a) original mesh, (b) after remeshing (according to the lines of curvature). Reproduced from Alliez *et al.* 2003

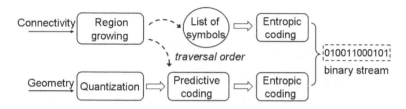

Figure 2.6   Generic scheme of single rate compression algorithms

which provides a list of symbols and defines a traversal order of the mesh, while the geometry is coded by quantization and then prediction guided by this traversal order. Figure 2.6 illustrates this generic scheme.

Connectivity coding

In a standard 3D mesh representation, the connectivity of the mesh is represented by cyclic lists of vertex indices, representing the faces (see Figure 1.4); this representation is extremely redundant. The first author to propose a method for connectivity encoding (suitable only for triangular meshes) was Deering (1995); his method consists of the definition of a vertex-ordered list such that each vertex of this list forms a triangle with the two preceding vertices. Thus, he cuts the mesh into *triangular strips* which lead to an effective connectivity encoding, provided that their lengths are sufficient. In a different way, Taubin and Rossignac (1998) proposed a method based on *spanning trees*. Their algorithm is grounded in the theoretical results of Turan (1984) who established that a planar graph can be encoded with a constant number of bits per vertex, using two covering trees: one for its faces and the other for its vertices.

Most of the recent methods are based on a **region growing** approach, which extends a region over the mesh and incrementally encodes the mesh elements and their incidence relations to the region. This class of algorithms can be separated into three main categories, according to which elements the description of the update is associated with: face (Gumhold and Strasser 1998; Kronrod and Gotsman 2001; Rossignac 1999; Szymczak *et al.* 2001), edge (Isenburg 2000; Isenburg and Snoeyink 2000) or vertex (Touma and Gotsman 1998). These algorithms can also be classified according to the type of mesh that they consider: triangular (Gumhold and Strasser 1998; Isenburg 2000; Rossignac 1999; Touma and Gotsman 1998) or polygonal (Isenburg and Snoeyink 2000; Kronrod and Gotsman 2001). These algorithms are based on several common components illustrated in Figure 2.7. The *region* describes

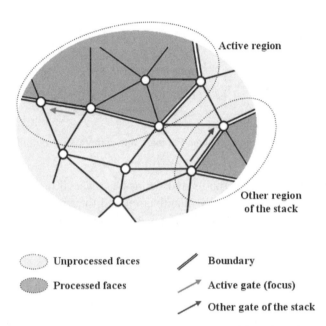

Figure 2.7 Principal elements of *region growing* algorithms based on connectivity encoding

the set of faces already *processed* by the algorithm. The edges which are adjacent to only one processed face comprise the *boundary* of the *region*. Among these *boundary edges*, the one which is adjacent to the next face to be processed is called the *focus* (or the *gate*). The notion of *edge* that we use here describes in fact an *oriented edge*, a standard edge of the mesh being composed of two *oriented edges* of opposite directions. Two *oriented edges* are said to be *adjacent* if they share the same vertices but are oriented differently. The growing operations are described by *symbols*. Once the object is entirely covered by the *growing region*, the list of created *symbols* permits retrieval of the connectivity of the mesh (by a reverse process, at the decoding).

The region can split during the growing process, hence several regions will grow over the mesh. Their processing sequence is managed through a stack. Thus there is in fact a stack of regions, a stack of boundaries and a stack of gates. Among these groups, the current elements are called the *active region*, *active boundary* and *active gate*.

Figure 2.8 illustrates the encoding algorithm *FaceFixer* from Isenburg and Snoeyink (2000), an edge-based scheme. With this approach the connectivity of the mesh is encoded by a list of *n* symbols, among an alphabet of length

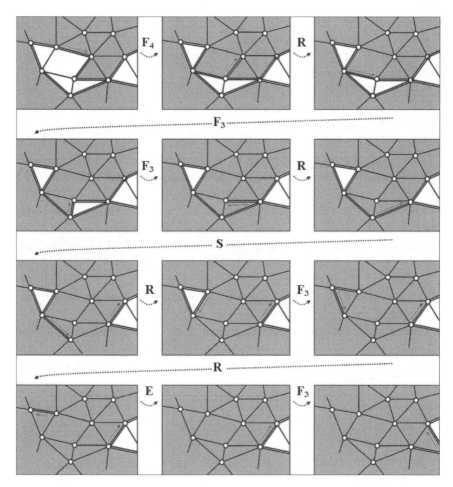

Figure 2.8 Example of the connectivity encoding algorithm *FaceFixer* from Isenburg and Snoeyink (2000)

$k(k \approx 10)$ with n the number of edges of the mesh. The symbols are generated according to the adjacent elements of the active gate. The algorithm distinguishes seven different situations respectively, represented by the symbols F, R, L, S, E, H and M, which are detailed in the next sections (and illustrated in Figure 2.8):

- *Symbol F_d*: The active gate is adjacent to an *unprocessed* face, of degree d. The active boundary is thus extended around this face. The symbol is associated with the degree d of the face.

- *Symbol R*: The active gate is adjacent to the next oriented edge of the active boundary (which connects the same two vertices but is oriented differently). They create a fold which is eliminated; the active gate is shifted on the preceding boundary edge.
- *Symbol L*: The active gate is adjacent to the previous oriented edge of the active boundary (which connects the same two vertices but is oriented differently), and is thus shifted on the next boundary edge.
- *Symbol S*: The active gate is adjacent to another oriented edge of the active boundary, which is not the previous nor the next one. The active boundary is split and thus a new boundary and a new gate are added to the respective stacks.
- *Symbol E*: The active gate is adjacent to another oriented edge of the active boundary, which is both the previous and the next one. The active boundary, which contains only these opposite edges, is eliminated. The encoding process continues with another boundary from the stack.
- *Symbol H_s*: The active gate is adjacent to a hole of size s (number of edges around the hole), thus the active boundary is extended over it. The symbol is associated with the size s of the hole.
- *Symbol M_{ijk}*: The active gate is adjacent to an oriented edge from another boundary from the stack, hence these boundaries are merged together. The symbol is associated with three integers representing the context and the position of the merging operation.

This algorithm produces n symbols, with n the number of edges of the mesh. For each face, there is one F, for each hole there is one H and for each handle (genus $\neq 0$) there is one M. This encoding process also defines a traversal order in the mesh.

The decoding process reconstructs the connectivity by processing the symbols in reverse order. A decoding example is illustrated in Figure 2.9. At each addition of a new vertex, its position is known since the traversal order of the vertices was fixed at the encoding process.

Once symbols have been created, the objective is to convert them into a binary stream as small as possible with entropic coders (see Section 2.3.1). Figure 2.10 illustrates the symbol repartition for two sample 3D objects: *Fandisk* (6495 vertices, 19 479 edges) and *Swivel* (2743 vertices, 8229 edges). In both cases, symbols R and F_3 are predominant. This highly non-uniform repartition is particularly adapted to entropic coding. For example, after

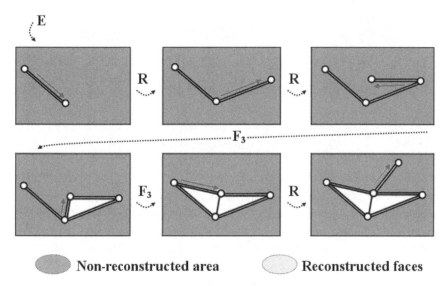

Figure 2.9 Example of decoding process associated with the FaceFixer algorithm from Isenburg and Snoeyink (2000)

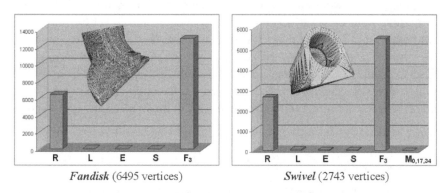

Fandisk (6495 vertices)   *Swivel* (2743 vertices)

Figure 2.10 Symbol distributions produced by the *FaceFixer* connectivity encoding of the 3D models *Fandisk* (6495 vertices, 19 479 edges) and *Swivel* (2743 vertices, 8229 edges)

arithmetic encoding, the results are respectively 0.76 and 0.84 bits per symbol for *Fandisk* and *Swivel* or 2.28 and 2.53 bits per vertex.

Figure 2.11 presents the algorithm from Touma and Gotsman (1998) which is a vertex-based scheme. In this algorithm, the vertex pointed by the *focus*

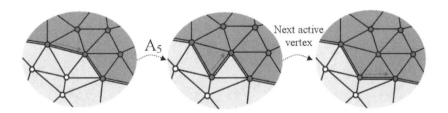

Figure 2.11   The connectivity coding scheme from Touma and Gotsman (1998)

(the gray arrow) is called the *active vertex*. At each iteration, the focus turns around this vertex (anticlockwise direction) and iteratively encodes the valence of adjacent vertices with their integration in the region. At the first iteration presented in Figure 2.11 the integrated vertex (its color turns from white to gray) has a valence of 5; the associated code is thus A_5. With the second iteration, all the neighbors of the active vertex have been visited, thus the algorithm considers a new active vertex and updates the region and the focus.

The region growing approach from Touma and Gotsman (1998) (adapted for triangular meshes) actually encodes the valence of each vertex, plus a limited number of exception codes. Alliez and Desbrun (2001b) have improved this algorithm by limiting, thanks to some heuristics, the occurrence of these exception codes. These approaches can be regarded as a new class of algorithms, the *valence-based approaches*. Recently, Khodakovsky *et al.* (2002) and Isenburg (2002) have generalized them to arbitrary polygonal meshes.

Geometric coding

Much less work has been done on geometry compression, which represents the largest part of the compressed stream. Most of the existing algorithms follow a similar scheme: first quantization, then prediction and finally entropic coding.

The coordinates of the vertices are first quantized (generally on 10 or 12 bits); the simplest method is the uniform quantization (Deering 1995; Touma and Gotsman 1998). Other more sophisticated methods have been developed. Chow (1997) segmented the mesh according to the curvature, and then applied a different quantization step to each region. Lee and Ko (2000) used an algorithm based on vector quantization. Lastly, Sorkine *et al.* (2003)

concentrated the quantization error in the low frequencies of the spectrum of the object, working on the assumption that the human eye is more sensitive to the variation of the normals than to the geometrical error.

Quantized coordinates are not directly encoded. Firstly the dynamic range of the values is reduced using geometric prediction techniques based on the traversal order dictated by the coding of the connectivity. These techniques are numerous, the simplest being differential coding (Deering 1995): a vertex is encoded by its difference vector to the previous one. For example, if we consider the vertex traversal order produced from the connectivity encoding of the Fandisk 3D model by the FaceFixer algorithm presented in the previous section, these differential vectors have a highly localized distribution. If we consider a 10-bit quantization, the corresponding differential coordinates are then included in the interval $]-2^{10}, 2^{10}[$, thus $]-1024, 1024[$, but 95% are in $]-30, 30[$! Hence the arithmetic encoding of these differential vectors is quite efficient: respectively 5.13, 5.26 and 4.10 bits per coordinate for X,Y and Z, and thus 14.49 bits per vertex instead of 30.

More recently Touma and Gotsman (1998) introduced the efficient *parallelogram prediction*. Figure 2.12(a) illustrates this scheme. To encode the vertex V_4, the authors consider the triangle V_1, V_2, V_3 already coded, and suppose that the polygon V_1, V_2, V_3, V_4 defines a parallelogram, thus they build the vertex V_4'. Then the vertex V_4 is encoded only by its difference vector $\vec{d}$ with V_4'. Isenburg and Alliez (2002) generalized this technique, initially adapted for triangular meshes, for arbitrary polygonal meshes. The drawback of these methods is that they cannot be optimal since the order of enumeration of the vertices is dictated by the coding of the connectivity. Thus, in the algorithm of Kronrod and Gotsman (2002) this is the coding of the geometry (by *prediction trees*) which defines the traversal of the mesh. Finally, we quote the *Angle-Analyzer* algorithm from Lee *et al.* (2002), illustrated

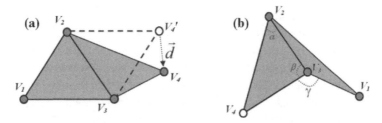

Figure 2.12 Illustration of two geometric prediction algorithms: (a) parallelogram prediction (Touma and Gotsman 1998), (b) Angle-Analyzer (Lee *et al.* 2002)

in Figure 2.12(b). Knowing that the triangle V_1, V_2, V_3 has already been encoded, the vertex V_4 is encoded by the three angles α, β and γ. The largely non-uniform and localized distribution of these angles is particularly adapted to entropic coding.

Gigantic mesh compression

Due to the improvement in scanning technologies, very large polygonal meshes are now available. For instance, some 3D models represent several gigabytes of data (Levoy *et al.* 2000) (hundreds of millions of triangles). These massive data sets require specific algorithms, since the whole model and the data structure required for standard compression algorithms do not fit in normal memory. Ho *et al.* (2001) proposed cutting meshes into smaller pieces (manageable by standard algorithms) which are then compressed separately. In a different way, Isenburg and Gumhold (2003) proposed using an external memory data structure for compression: the mesh is implicitly cut but into a much larger number of smaller pieces; at run-time only a small number of pieces are kept in memory, and the majority remains on disk. Cignoni *et al.* (2003) proposed a similar approach for mesh simplification. Another approach is to eliminate indices associated with the mesh elements and to consider triangle soup, which streams from the disk to the memory triangle per triangle (Cignoni *et al.* 2004; Lindstrom 2000; Wu and Kobbelt 2003). Finally Isenburg and Lindstrom (2005) introduced a *streaming* format for polygonal meshes, based on coherent and compatible ordering of the mesh vertices and polygons. These *streaming meshes* are then used for compression (Isenburg *et al.* 2005) and simplification (Isenburg *et al.* 2003).

Recent advances

We have seen that remeshing can be considered as compression inasmuch as it yields a decrease in the vertex/face number and therefore the amount of information to be encoded. Most of the algorithms for connectivity and geometric encoding are particularly efficient in the cases of uniform and regular meshes, leading several authors to propose remeshing methods for producing such meshes. Szymczak *et al.* (2003) transformed the input mesh into a piecewise regular one, which they encoded using an adapted version (Szymczak *et al.* 2001) of the *EdgeBreaker* algorithm from Rossignac (1999). Gu *et al.* (2002) proposed a completely regular remeshing technique which maps the mesh on a rectangular grid (*geometric image*), which can then be encoded using 2D image compression algorithms.

On the whole compression rates of the single rate algorithms presented in this section are between 10 and 16 bits per vertex (for a 10-bit geometric quantization). As regards connectivity, on which a lot of work has been done in the last 10 years, valence-based approaches (Alliez and Desbrun 2001b; Isenburg 2002; Khodakovsky *et al.* 2002; Touma and Gotsman 1998) provide the better results (about 3 bits per vertex) and get close to the graph encoding lower bound. In most of the approaches, geometric encoding is dictated by the connectivity symbols, and is therefore not optimal. Thus some recent approaches have given priority to geometry (Kronrod and Gotsman 2002; Sorkine *et al.* 2003). Finally, recent remeshing-based approaches, particularly geometric images, provide excellent results.

Progressive compression

Progressive (or multiresolution) approaches are linked to the notion of refinement. The main idea is to transmit a simple coarse mesh (low resolution) and a refinement sequence permitting an incremental reconstruction of the mesh during the transmission. Progressive compression algorithms are strongly linked with incremental decimation (see section 2.4.1). Indeed most of them consist of decimating the mesh (vertex/edge suppressions) while storing the information necessary for process inversion, i.e. the refinement (vertex/edge insertions during the transmission). The existing approaches differ in the way they decimate the mesh and store the refinement information (where and how can this be refined?).

Progressive meshes

The concept of *progressive mesh* was introduced by Hoppe (1996). During the encoding, the mesh is simplified iteratively by a sequence of edge collapses. At the decoding it is refined, vertex by vertex, by the dual operator: vertex split (see Figure 2.4). This technique, generalized by Popovic and Hoppe (1997) for arbitrary non-manifold meshes, has the advantage of having a fine granularity but remains expensive in memory (around 16 bits per vertex for connectivity), because the index number of each vertex to be split is encoded explicitly. Thus, Taubin *et al.* (1998) and Pajarola and Rossignac (2000) simplified the models, by combinations of edge collapses applied to several edges at the same time. Thus at each iteration, a whole region of the mesh is simplified. Therefore the granularity is coarser than for the algorithm of Hoppe (1996) but the cost of the encoding is reduced (around 10 and 7 bits per vertex for connectivity respectively for Taubin *et al.* (1998) and Pajarola and Rossignac (2000)). Similarly, Cohen-Or *et al.* (1999) applied the collapse/split operators to sets of independent vertices. Their approach was improved by Alliez and Desbrun (2001a) who considered in particular the vertex valences; their algorithm gets close to single rate schemes, in term of compression rate for connectivity (around 4 bits per vertex). Their approach is illustrated in Figure 2.13. For all the approaches presented in this section, the geometry is encoded in a similar way to the single rate schemes, by quantization and then prediction (each new vertex generated by vertex split can be effectively predicted thanks to its already transmitted neighbors).

A geometric approach

In a different way, Devillers and Gandoin (2000) proposed a progressive compression algorithm mainly driven by the geometry. They recursively subdivided the bounding box of the object into smaller and smaller cells, until they contain, at most, one vertex. At each subdivision level, they transmit only occurrences of points (the authors demonstrated that it is enough to reconstruct a point set). Concerning connectivity, they encoded vertex splits which occur at each subdivision step. The authors have generalized this approach, which gives excellent results, to arbitrary non-manifold meshes (Gandoin and Devillers 2002). An example of the progressive transmission is illustrated in Figure 2.14.

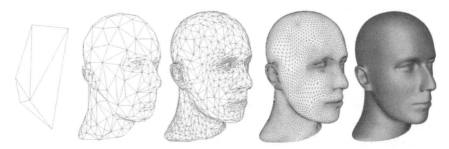

Figure 2.13 Progressive transmission, from the algorithm of Alliez and Desbrun. Reproduced from P Alliez and M Desbrun, 2001a

Figure 2.14 Progressive transmission, from the algorithm of Gandoin and Devillers (2002). Courtesy of P M Gandoin

Spectral methods

Several authors have exploited frequency analysis, which is extremely popular in 2D image compression, to develop new algorithm for 3D mesh progressive compression. Karni and Gotsman (2000) applied the spectral decomposition from Taubin (1995) by projecting the geometry of the mesh onto the eigenvectors of its Laplacian matrix. A geometrical frequency spectrum is obtained, quantized and then transmitted iteratively from low to high frequencies. An example of such decomposition is shown in Figure 2.15. Results are excellent, although the calculation of the eigenvalues of the Laplacian matrix is highly complex and is a strong limitation. The input mesh needs to be cut into smaller pieces before being processed. Moreover, the connectivity remains unchanged during the transmission; the progressiveness concerns only the geometry.

Wavelets

In the same way wavelets were introduced in the field of 2D image processing by Mallat (1989) for multiresolution decomposition and compression.

Figure 2.15 Example of spectral decomposition from Taubin (1995) of a 3D object. The object is shown after reconstruction with respectively the 5, 10, 60 and 500 lowest spectral coefficients (among 500)

Lounsbery *et al.* (1997) have defined a wavelet basis, based on the sub-division rules from Loop (1987) (see Section 1.1.3 for more details about subdivision surface), for the multiresolution decomposition of 3D triangular meshes. The decomposition to a lower resolution level is carried out by a mechanism of inverse subdivision. Khodakovsky *et al.* (2000) defined a progressive compression algorithm based on these wavelets, which gives better results than almost all classical methods (around 8 bits per vertex for geometry plus connectivity). The drawback is that the mesh to be decomposed must have a subdivision connectivity (i.e. semi-regular), and therefore often needs a remeshing. This is why Valette and Prost (2004a) have defined a wavelet basis for irregular meshes with an associated compression algorithm (Valette and Prost 2004b): their results are slightly lower than the algorithm of Khodakovsky *et al.* (2000) but the initial connectivity is preserved (no remeshing is needed). Figure 2.16 illustrates several levels of resolution of the mesh Fandisk decomposed on this wavelet basis.

Recent advances

The most recent works, whose results are the most impressive, are based on a previous remeshing (like the wavelet-based scheme of Khodakovsky

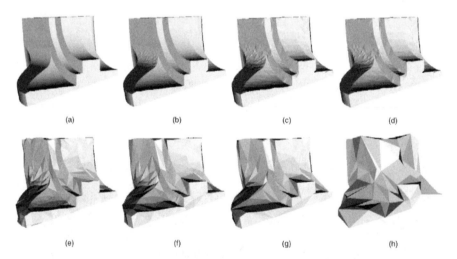

Figure 2.16 (a) Original mesh (6475 vertices). (b–h) Lower resolution levels, wavelet basis from Valette and Prost (2004a) (respectively 2438, 1250, 648, 378, 230, 155 and 83 vertices). Reproduced by permission of IEEE

et al. (2000)) or on a different modeling. The concept of *normal mesh* was introduced by Guskov *et al.* (2000). It is a multiresolution representation of a mesh, in which each level can be written as a normal offset from a coarser version, thus the geometry is represented by only one scalar instead of three. Khodakovsky and Guskov (2003) introduced a wavelet-based coding scheme especially adapted for this model. Praun and Hoppe (2003) defined an algorithm of spherical parameterization allowing a 3D mesh to be mapped onto a rectangular grid: the *spherical geometry image*. This model, an extension of the *geometry image* of Gu *et al.* (2002), is then encoded by the authors (Hoppe and Praun 2005) using 2D image progressive compression techniques such as wavelets. Results are impressive, and are illustrated in Figure 2.17. Peyré and Mallat (2005) further improved the compression rate by using bandelets instead of wavelets.

The high number and diversity of the progressive compression methods presented here reflect the dynamism of the scientific community on this topic. Whereas the constraint of progressiveness involved a strong reduction in the effectiveness of coding in the first approaches (Hoppe 1996), the more recent algorithm from Alliez and Desbrun (2001a) reached the compression rates obtained by single rate algorithms. The approaches based on wavelets, like those of Khodakovsky *et al.* (2000) or Valette and Prost

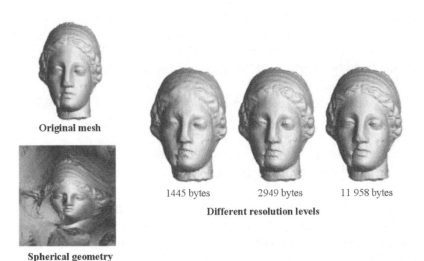

Original mesh

Spherical geometry
image

1445 bytes   2949 bytes   11 958 bytes
Different resolution levels

Figure 2.17 Progressive compression based on *spherical geometry images* (Hoppe and Praun 2005). Reproduced with kind permission of Springer Science and Business Media

(2004b), provide even better results, the principal issue being to provide the most regular possible mesh in order to improve the performance. As for single rate compression, the actual issue is no longer mesh encoding but rather shape encoding, for which the connectivity conservation is not essential. The most recent and most impressive approaches are based on a previous remeshing or parameterization (Hoppe and Praun 2005; Khodakovsky and Guskov 2003; Khodakovsky *et al.* 2000; Peyré and Mallat 2005) and these pre processing steps constitute the real issue because they drive the efficiency of the compression algorithms.

2.4.2   Other 3D Model Compression

Parametric surfaces

Compression of parametric surfaces is a field that has not been much investigated by the scientific community. Santa-Cruz and Ebrahimi (2002) presented a compression scheme for NURBS surfaces, where node vectors and control points are encoded by prediction and entropic coding, using an arithmetic coder. Furukawa and Masuda (2002) proposed a quite different algorithm where the boundary curves are extracted from the NURBS surface and then, starting from these curves, an interpolation surface is calculated: a Coons surface (Coons 1964). Differences between the interpolated surface and the original one are then extracted in the form of two distance maps representing two precision levels. The original NURBS surface is finally encoded by using only the boundary curves and the displacement maps. This idea of distance map is also considered by Krishnamurthy and Levoy (1996) who associated to the B-spline patches, coming from their approximation algorithm, some distance maps, in a scalar or vectorial form to represent details. Lastly, Gopi and Manocha (1999) proposed a compression technique based on an approximation of the B-spline model in terms of cubic triangular Bézier patches. Their method enables, moreover, the generation of several levels of detail (or resolution levels).

This idea of multiresolution decomposition is strongly related to compression. Forsey and Bartels (1988) defined a multiresolution B-spline model. From an original B-spline surface they created coarser versions (by dividing the number of nodes and control points by 2 with each iteration), which they optimized in terms of least squares. Other hierarchical decomposition approaches are based on wavelets: Finkelstein and Salesin (1994) defined a wavelet basis for the decomposition of uniform cubic B-spline curves; their

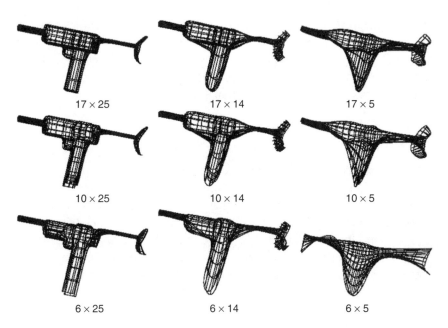

Figure 2.18   Wavelet decomposition of a bi quadratic B-spline surface (17 × 25 control points) using the algorithm of Kazinnik and Elber (1997). Reproduced with permission of Blackwell Publishing

work was generalized for surfaces by Martin (1995). Lyche and Morken (1992) introduced a wavelet basis associated with NURBS basis functions. These wavelets were applied by Kazinnik and Elber (1997) for the decomposition of NURBS surfaces. Their algorithm is illustrated in Figure 2.18. Masuda *et al.* (1999) used a similar approach but considered the DCT (Discrete Cosine Transform) decomposition and coding of NURBS surfaces.

Discrete model

Few lossless compression methods exist for discrete models. Fowler and Yagel (1994) generalized the *DPCM* (Differential Pulse Code Modulation) method for discrete 3D data, coupled to a *Huffman* coding (Huffman 1952). This algorithm provides a compression rate of only 50% compared with the original data. Lossy compression methods are numerous, and are mostly based on transforms, such as the DCT, applied notably by Yeo and Liu (1995). Other approaches use Vectorial quantization (Ning and Hesselink

1992) or fractals (Cochran *et al.* 1996). Another class of algorithms considers a hierarchical decomposition, either Laplacian (Ghavamnia and Yang 1995) or by wavelets (Ihm and Park 1998; Muraki 1993). Recently Dupont *et al.* (2003) proposed a progressive compression method based on the union of balls and cones from a medial axis transformation.

2.5 Compression Based on Approximation

Some representation schemes are much more compact than others, in terms of amount of data. For example, for a given shape a parametric or subdivision surface modeling will provide a much lighter model than a triangulation in a polygonal mesh. Thus there exist a lot of algorithms which take as input polygonal meshes or point clouds and provide lighter approximating surfaces, like parametric, subdivision, implicit, or primitive-based ones. These approximations can be considered as compression schemes since they provide a very compact version of the same shape. This chapter presents the existing approximation methods.

2.5.1 Parametric Surface Approximation

The problem of the approximation of a 3D mesh or a 3D point cloud, by a patchwork of B-spline or NURBS surfaces, has been tackled by many authors. A state of the art of various techniques was written by Dierckx (1993). However, the majority of the existing techniques consider a simple topology or require human intervention to define the network of patches. Hoschek *et al.* (1989), Rogers and Fog (1989), Sarkar and Menq (1991) and Ma and Kruth (1995) regarded the surface as made of only one B-spline patch. Similarly Forsey and Bartels (1995) considered the fitting of only one hierarchical B-spline surface on a rectangular grid. Milroy *et al.* (1995) and Krishnamurthy and Levoy (1996) implemented the approximation of an arbitrary topology 3D object by a network of B-spline patches, but these methods require human intervention to define the boundaries of the patches. This user intervention is avoided by Eck and Hoppe (1996) who simplified the input mesh into a coarse mesh that they used as a control mesh to generate B-spline surfaces using the algorithm of Peters (1994). Their algorithm is illustrated in Figure 2.19. These B-spline approximation approaches remain quite complex, indeed a correct network of patches (i.e. with a correct topology) is difficult to define, and the continuity at the joins between the patches

Figure 2.19 Illustration of the B-spline approximation algorithm of Eck and Hoppe (1996): left, original mesh (69 473 faces); middle and right, B-spline surfaces (respectively 72 and 153 patches). Reproduced from ACM Siggraph

is complicated to manage. For these reasons these approximation methods based on B-spline and NURBS are used particularly in the fields of reconstruction and reverse engineering, and are more and more dropped in favor of subdivision surface-based algorithms (see the following section) which avoid these topology and continuity problems. Moreover, several methods build a network of B-spline patches starting from a subdivision control polyhedron (Peters 1994, 1995, 2000).

2.5.2 Implicit Surface Approximation

Many authors have investigated the problem of approximating a 3D mesh or a 3D point cloud with algebraic implicit surfaces (Ahn *et al.* 2002; Keren and Gotsman 1999; Keren *et al.* 1994; Taubin *et al.* 1994). These aim at finding polynomial coefficients which minimize the distance from the implicit surface to the 3D data. These algorithms are based on a non linear optimization generally solved by computation-heavy algorithms like least squares (Taubin *et al.* 1994). The approach of Keren and Gotsman (1999) is quite different, in that they consider only certain families of polynomials, which satisfy several good properties, in their approximation algorithm. That often led to the use of polynomials of degrees higher than necessary, but avoids many robustness and stability problems. Lastly, Lei and Cooper (1998) and Blane *et al.* (2000) presented linear algorithms that were more stable and rapid than traditional non linear ones. Recently Sahin and Unel (2005) introduced a linear algorithm that even improves stability for high-degree surfaces.

Even if these fitting algorithms give quite good results, the fact of using algebraic surfaces to represent arbitrary geometrical objects presents several drawbacks: the representational power of algebraic surfaces of low degree is often insufficient, and if we increase the degree of the polynomials, associated surfaces become very complex to handle and the fitting procedure involves major calculation and stability problems. That is why most of the current implicit surface fitting algorithms rather concern non-algebraic surfaces. One of the first authors to consider this model for the approximation/reconstruction issue was Muraki (1991) who approximated 3D range data with a set of *blobs*. Blob number and sizes are calculated using a non linear energy minimization. Savchenko *et al.* (1995) presented a similar algorithm but considered particles associated with a different potential function. Turk and O'Brien (1999) applied a similar technique for 3D shape deformation. These algorithms are limited to small objects, because of the high processing time due to the global non linear optimization. Several authors have tried to resolve this complexity problem. Yngve and Turk (2002) first converted the input 3D data (polygonal meshes) into a volumetric representation (voxels in a regular grid), but the major drawback is the loss of fine detail. Morse *et al.* (2001) used compact support potential functions which largely reduce processing time and necessary memory. Carr *et al.* (2001) first evaluated the implicit surface on a subset of points and then repeatedly refined it by considering more and more points until the desired precision was obtained. Their fitting method is illustrated in Figure 2.20. Lastly, Li *et al.* (2004) applied

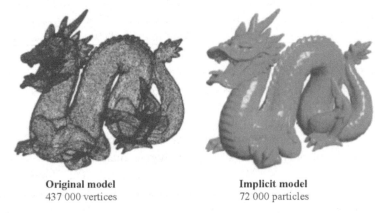

Original model   Implicit model
437 000 vertices   72 000 particles

Figure 2.20   Original mesh and approximating particle-based implicit surface, fitting algorithm of Carr *et al.* (2001). Courtesy of Applied Research Associates NZ Ltd

ellipsoidal constraints on their implicit surfaces so as to simplify calculations. More recently, Samozino *et al.* (2006) considered the Voronoi diagram of the input data points to improve the fitting, while Kanai *et al.* (2006) introduced a hierarchical technique based on a novel error function.

2.5.3   Subdivision Surface Approximation

Several methods exist for subdivision surface fitting, most of them taking as input a dense mesh, simplifying it to obtain a base coarse control mesh and then displacing the control points (geometry optimization) to fit the target surface. Lee *et al.* (2000), Ma *et al.* (2004), Mongkolnam *et al.* (2003) and Marinov and Kobbelt (2005) used the quadric error metrics from Garland and Heckbert (1997) for simplification. Kanai (2001) used a similar decimation algorithm which directly minimizes the error between the original mesh and the subdivided simplified mesh; his method is illustrated in Figure 2.21.

With these simplification-based approaches, the control mesh connectivity strongly depends on the input mesh. The fitting method of Susuki (1999) remains independent of the target mesh connectivity, by iteratively subdividing and shrinking an initial hand-defined control mesh toward the target surface. Jeong and Kim (2002) used a similar shrink wrapping approach. In the same way, Cheng *et al.* (2004) constructed an octree partition of the target surface and then triangulated it using the marching cube algorithm. Lavoué *et al.* (2007) used boundaries and curvature information from segmented regions to construct a topologically correct control mesh while remaining independent of the original connectivity. Lastly, Li *et al.* (2006) considered a

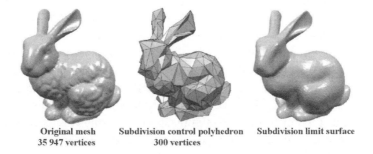

Original mesh   Subdivision control polyhedron   Subdivision limit surface
35 947 vertices   300 vertices

Figure 2.21   Subdivision surface fitting example, algorithm of Kanai (2001)

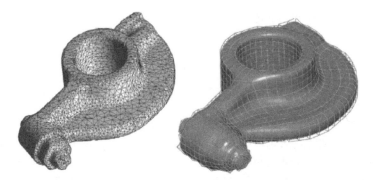

Figure 2.22 Original 3D mesh ($\approx$23 000 vertices) and associated control mesh (2021 control points) and limit surface, from the algorithm of Li *et al.* (2006). (c) Eurographics Association 2006. Reproduced by kind permission of the Eurographics Association

global parameterization and a quad dominant remeshing algorithm (Ray *et al.* 2006) to create an appropriate control mesh. Their approach is illustrated in Figure 2.22. From a polygonal 3D mesh they obtained a T-spline representation (Sederberg *et al.* 2003), which is basically similar to a Catmull–Clark subdivision surface.

Concerning the geometry optimization, Lee *et al.* (2000) and Hoppe *et al.* (1994) sampled a set of points from the original mesh and minimized a quadratic error to the subdivision surface. This technique was recently improved by Marinov and Kobbelt (2005) which introduced parameter corrections. Susuki (1999) proposed a local and faster approach, also used by Jeong and Kim (2002) and Mongkolnam *et al.* (2003). The positions of the control points are optimized, only by reducing the distance between their limit positions and the target surface. Hence only subsets of the surfaces are involved in the fitting procedure, thus results are not so precise. Litke *et al.* (2001) also introduced a local algorithm, based on *quasi-interpolation*, to compute detail coefficients on a Catmull–Clark surface. Ma *et al.* (2004) considered the minimization of the distances from vertices of the subdivision surface after several refinements to the target mesh. Lavoué *et al.* (2007), Cheng *et al.* (2004) and Marinov and Kobbelt (2005) used a similar algorithm. However they have not considered a point-to-point distance minimization but a point-to-surface minimization, by using the local quadratic approximants introduced by Pottmann and Leopoldseder (2003).

Concerning connectivity optimization, Hoppe *et al.* (1994) optimized the connectivity by trying to collapse, split or swap each edge of the control polyhedron. Their algorithm produces high-quality models but needs of course extensive computing time. Recently, Marinov and Kobbelt (2005) subdivided faces associated with high errors and flipped some edges to regularize vertex valences, similar to Cheng *et al.* (2004). Lavoué *et al.* (2007) and Li *et al.* (2006) optimized the connectivity of the control mesh by analyzing curvature directions of the target surface.

Figure 2.23 illustrates the subdivision surface fitting algorithm from Lavoué *et al.* (2007). The algorithm begins with a decomposition of the object into surface patches. Then the main idea is to approximate first the region's boundaries and then the interior data. Thus, for each patch, a first step approximates the boundaries with subdivision curves (associated with control polygons) and creates an initial subdivision surface by linking the boundary control points with respect to the lines of curvature of the target surface. Then, a second step optimizes the initial subdivision surface by iteratively moving control points and enriching regions according to the error distribution. The final control mesh defining the whole model is then created assembling all the local subdivision control meshes. This algorithm preserves sharp features by marking the corresponding control edges (the lighter lines in Figure 2.23(d)).

2.5.4 Superquadric-based Approximation

Several methods consider only one primitive to approximate a set of 3D data, as is the case of Solina and Bajcsy (1990) and Terzopoulos and Metaxas (1991). Others consider a set of primitives; 3D data are thus segmented and then modeled by a graph of superquadrics. This class of approaches is

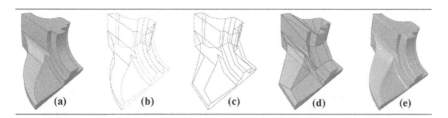

Figure 2.23 The different steps of the surface fitting scheme from Lavoué *et al.* (2007) for the Fandisk object (6495 vertices): (a) segmentation, (b) boundary extraction, (c) boundary approximation, (d) subdivision control mesh (75 control points), (e) limit surface. Reproduced with permission of Blackwell Publishing

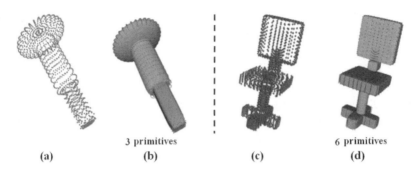

Figure 2.24 Segmentation and approximation of a 3D point cloud by a set of superellipsoids, algorithm of Chevalier *et al.* (2001): (a,c) original 3D data, (b,d) approximating superellipsoids. Reproduced from L Chevalier, 2001

particularly relevant within the framework of 3D object indexing or recognition. The first author to follow this tendency was Pentland (1990). Gupta and Bajcsy (1993) and Leonardis *et al.* (1997) segmented and approximated range images by a set of superellipsoids. More recently, Chevalier *et al.* (2001) segmented and approximated a 3D point cloud by a set of superellipsoids, by a *split and merge* technique. Their approach is illustrated in Figure 2.24.

2.6 Normative Aspect: MPEG-4

MPEG-4 is a standard that enables the representation, composition and transmission of multiple multimedia objects on a terminal. These objects were mostly image, video and sound. However, with the increasing number of applications around 3D data (animation, games, CAD, scientific visualization), the MPEG-4 standard has integrated 3D graphics capabilities within the animation framework extension (AFX) (MPEG-4 2002a,2002b). This standard mostly relies on VRML 2.0 with the addition of new functionalities, in particular compression and streaming capabilities. This 3D graphics extension contains tools for texture mapping and animation of 3D content; the standard also describes many concepts and mechanism for shape representation and compression, as briefly described in the following sections.

Several surface representations are available, with associated compression techniques (Bourges-Sévenier and Jang 2004; Mignot and Garneau 2004):

- **Polygonal mesh**: This model was considered in the first 3D international standards (VRML97), but MPEG-4 provides several new features within the 3D mesh coding tool (3DMC). It offers a single resolution compression algorithm based on *topological surgery* (Taubin and Rossignac 1998), differential quantization and entropic coding. A multi resolution decomposition tool is also provided, based on the on the *progressive forest split* algorithm (Taubin *et al*. 1998). These tools support both manifold and non-manifold meshes.

- **Surface-based models**: The AFX supports NURBS, implicit and subdivision surfaces and provides simple associated compression tools. Loop and Catmull–Clark's subdivision rules are considered, together with extensions such as normal control and edge sharpness. Multi resolution subdivision surfaces are also supported, which enable details to be added at each subdivision iteration. This concept is strongly related to the 3D wavelets and thus permits support of the mesh wavelet compression.
- **Mesh grid representation (Salomie *et al*. 2004)**: This model enables a compressed and hierarchical representation of a 3D object. It combines a description of the 'global connectivity' between the vertices on the object's surface with a regular 3D grid of points. It enables regions of interest and levels of refinement to be defined for proper adaptation of the mesh resolution and precision in the viewing conditions.

2.7 Conclusion

In this chapter we have presented existing 3D compression techniques and more particularly polygonal mesh compression algorithms. The research community has been particularly active in this field for the last 10 years. Mono-resolution techniques now provides compression rates between 10 and 16 bits per vertex. First techniques have rather focused on connectivity, but more recent approaches give priority to geometry that represents the major part of the data. Concerning multi resolution techniques, whereas the constraint of progressiveness involved a strong reduction of the effectiveness in the first approaches, recent algorithms reach the compression rates obtained by mono-resolution ones. Most recent and effective algorithms focus more on compressing the shape of the object than the mesh itself. They carefully remesh the object (regularly or according to feature lines, for instance) so as to improve the performances of the further compression schemes. Indeed, the quality of the mesh represents a critical issue since it drives the efficiency of the compression. In the same manner another efficient way to obtain high compression rates is to change the whole representation of the object using approximation schemes. For example, approximating a dense polygonal mesh with an implicit or subdivision surface can lead to an extremely high compression rate since these models are far more compact. However, this class of algorithm is quite complex to handle.

2.8 Summary

Compared with audio, image or even video, 3D data may often represent a quite gigantic size in memory. In this context the need for efficient tools to reduce the size of this 3D content, mostly represented by polygonal meshes, becomes even more acute, particularly to reduce the transmission time for low-bandwidth applications. That is precisely the objective of compression techniques.

A compression algorithm often consists of two main parts:

- Finding a concise/synthetic representation of the 3D model, even if it implies some loss of precision or some errors regarding the original object.
- Suppressing the redundancy using classical coding techniques, such as quantization, prediction and entropic encoding.

The most widespread representation for 3D data is the polygonal mesh. It consists of a set of vertices and facets that describe the surface of the 3D object. It implies a large amount of data: the 3D coordinates of the vertices and the connectivity (the way the vertices are linked together) have to be encoded. Hence many compression algorithms have been developed by the scientific community in the last decade. They can be separated into three main categories:

- **Simplification**: These algorithms aim to transform the input mesh into a new mesh containing less faces, vertices and edges, but representing the same shape. This is not compression in the strict sense of the word, but rather data reduction.
- **Single rate compression**: The objective is to suppress the redundancy of the original data. In this class of algorithm, the whole 3D mesh is decoded and reconstructed at the end of the transmission.
- **Progressive compression**: During the decoding, the 3D mesh is reconstructed incrementally, providing access to intermediate states of the object. These progressive (or multiresolution) algorithms first transmit a low-resolution version of the mesh, followed by iterative refinement information which allows construction of finer and finer versions.

New trends tend to consider algorithms that carefully remesh the object (semi-regularly, completely regularly, following the feature lines, etc.) so as to improve the efficiency of the further compression scheme.

Some compression algorithms exist also for the other existing 3D representations (NURBS, discrete model). However, the research community is far less productive for these models than for polygonal meshes.

The last technique for compressing 3D data, and particularly polygonal meshes, is to change the representation. It consists of approximating an input mesh with a more compact model like the NURBS, implicit or subdivision surface, for instance. This class of techniques can bring a real gain in compression, but the corresponding algorithms are often quite complex and time consuming.

2.9 Questions and Problems

2.9.1 Encoding and Quantization

1. What is the goal of entropic coding?
2. The geometry of a 3D mesh is quantized on a $500 \times 1000 \times 2000$ 3D grid. How many bits are necessary to encode respectively coordinates X, Y and Z of each vertex? Hence how many bits are necessary to encode the whole geometry of each vertex?
   Without quantization we assume that each coordinate is encoded by 4-byte floating numbers. What is the compression rate produced by the proposed quantization?

2.9.2 Mesh Compression/Approximation

1. Explain the difference between single rate compression and simplification.
2. What are the three main techniques for mesh simplification?
3. Which information represents the larger part of an encoded polygonal mesh: connectivity or geometry?
4. What is the benefit of progressive compression over single rate compression?
5. What is the benefit of approximating a 3D mesh with a subdivision surface?
6. Provide the symbols from connectivity encoding of the polygonal region in Figure 2.25 (white part) using the FaceFixer algorithm of Isenburg and Snoeyink (2000).

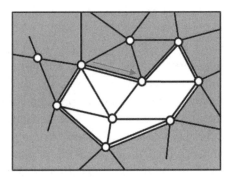

Figure 2.25 What are the symbols generated by the connectivity encoding of the white region, using the FaceFixer algorithm?

References

Ahn SJ, Rauh W, Cho HS and Warnecke HJ 2002 Orthogonal distance fitting of implicit curves and surfaces. *IEEE Transactions on Pattern Analysis and Machine Intelligence* **24**(5), 620–638.

Alliez P and Desbrun M 2001a Progressive encoding for lossless transmission of 3D meshes. *ACM SIGGRAPH*, pp. 198–205.

Alliez P and Desbrun M 2001b Valence-driven connectivity encoding of 3D meshes. *Computer Graphics Forum* **20**(3), 480–489.

Alliez P and Gotsman C 2005 Recent advances in compression of 3D meshes. In *Advances in Multiresolution for Geometric Modelling* (ed. Dodgson N, Floater M and Sabin M), pp. 3–26. Springer-Verlag.

Alliez P, Meyer M and Desbrun M 2002 Interactive geometry remeshing. *ACM Transactions on Graphics* **21**(3), 347–354.

Alliez P, Cohen-Steiner D, Devillers O, Levy B and Desbrun M 2003 Anisotropic polygonal remeshing. *ACM Transactions on Graphics* **22**(3), 485–493.

Blane MM, Lei Z, Civi H and Cooper DB 2000 The 3L algorithm for fitting implicit polynomial curves and surfaces to data. *IEEE Transactions on Pattern Analysis and Machine Intelligence* **22**(3), 298–313.

Boier-Martin I, Rushmeier H and Jin J 2004 Parameterization of triangle meshes over quadrilateral domains. *Eurographics Symposium on Geometry Processing*, pp. 197–208.

Bourges-Sévenier M and Jang ES 2004 An introduction to the MPEG-4 animation framework extension. *IEEE Transactions on Circuits and Systems for Video Technology* **14**(7), 928–936.

Carr J, Beatson RK, Cherrie JB, Mitchell TJ, Fright WR, McCallum BC and Evans TR 2001 Reconstruction and representation of 3D objects with radial basis functions. *ACM SIGGRAPH*, pp. 67–76.

Cheng KSD, Wang W, Qin H, Wong KYK, Yang HP and Liu Y 2004 Fitting subdivision surfaces to unorganized point data using SDM. *IEEE Pacific Graphics*, pp. 16–24.

Chevalier L, Jaillet F and Baskurt A 2001 Coding shapes with superquadrics. *IEEE International Conference on Image Processing*.

Chow M 1997 Optimized geometry compression for realtime rendering. *IEEE Visualization*, pp. 347–354.

Cignoni P, Montani C, Rocchini C and Scopigno R 2003 External memory management and simplification of huge meshes. *IEEE Transactions on Visualization and Computer Graphics* **9**(4), 525–537.

Cignoni P, Ganovelli F, Gobbetti E, Marton F, Ponchio F and Scopigno R 2004 Adaptive tetrapuzzles: efficient out-of-core construction and visualization of gigantic multiresolution polygonal models. *ACM SIGGRAPH*, pp. 796–803.

Cochran W, Hart J and Flynn P 1996 Fractal volume compression. *IEEE Transactions on Visualization and Computer Graphics* **2**(4), 313–322.

Cohen-Or D, Levin D and Remez O 1999 Progressive compression of arbitrary triangular meshes. *IEEE Visualization*, pp. 67–72.

Cohen-Steiner D, Alliez P and Desbrun M 2004 Variational shape approximation. *ACM Siggraph*, pp. 905–914.

Coons SA 1964 Surfaces for computer aided design of space figures. *MIT Preprint No. 299*.

Daly S 1993 The visible differences predictor: an algorithm for the assessment of image fidelity. In *Digital Images and Human Vision*, A. B. Watson, Ed. MIT Press, Cambridge, MA, 179–206.

Deering M 1995 Geometry compression. *ACM SIGGRAPH*, pp. 13–20.

Devillers O and Gandoin PM 2000 Geometric compression for interactive transmission. *IEEE Visualization*, pp. 319–326.

Dierckx P 1993 *Curve and surface fitting with splines*. Oxford University Press.

Dong S, Bremer PT, Garland M, Pascucci V and Hart JC 2006 Spectral surface quadrangulation. *ACM SIGGRAPH*.

Dupont F, Gilles B and Baskurt A 2003 Progressive transmission of 3D object based on balls and cones union from medial axis transformation. *IEEE International Conference on Image Processing*.

Eck M and Hoppe H 1996 Automatic reconstruction of B-spline surfaces of arbitrary topological type. *ACM SIGGRAPH*, pp. 325–334.

Echert M and Bradley A 1998 Perceptual quality metrics applied to still image compression *Signal Processing* **70**(3), 177–200.

Finkelstein A and Salesin DH 1994 Multiresolution curves. *ACM SIGGRAPH*, pp. 261–268.

Forsey D and Bartels R 1988 Hierarchical B-spline refinement. *Computer Graphics* **22**(3), 205–212.

Forsey D and Bartels R 1995 Surface fitting with hierarchical splines. *ACM Transactions on Graphics* **14**(2), 134–161.

Fowler J and Yagel R 1994 Lossless compression of volume data. *Symposium on Volume Visualization*, pp. 43–53.

Furukawa Y and Masuda H 2002 Compression of NURBS surfaces with error evaluation. *NICOGRAPH International*.

Gandoin PM and Devillers O 2002 Progressive lossless compression of arbitrary simplicial complexes. *ACM SIGGRAPH*, pp. 372–379.

Garland M and Heckbert PS 1997 Surface simplification using quadric error metrics. *ACM SIGGRAPH*, pp. 209–216.

Garland M, Willmott A and Heckbert P 2001 Hierarchical face clustering on polygonal surfaces. *ACM Symposium on Interactive 3D Graphics*, pp. 49–58.

Ghavamnia M and Yang X 1995 Direct rendering of Laplacian pyramid compressed volume data. *IEEE Visualization*, pp. 192–199.

Gopi M and Manocha D 1999 Simplifying spline models. *Computational Geometry* **14**(1–3), 67–90.

Gotsman C, Gumhold S and Kobbelt L 2002 Simplification and compression of 3D-meshes. In *Tutorials on multiresolution in geometric modeling* (ed. Iske A, Quak E and Floater M). Springer-Verlag.

Gu X, Gortler SJ and Hoppe H 2002 Geometry images. *ACM SIGGRAPH*, pp. 355–361.

Gumhold S and Strasser W 1998 Real time compression of triangle mesh connectivity. *ACM SIGGRAPH*, pp. 133–140.

Gupta A and Bajcsy R 1993 Volumetric segmentation of range images of 3D objects using superquadric models. *Computer Vision, Graphics, and Image Processing* **58**(3), 302–326.

Guskov I, Vidimce K, Sweldens W and Schroder P 2000 Normal meshes. *ACM SIGGRAPH*, pp. 95–102.

Ho J, Lee KC and Kriegman D 2001 Compressing large polygonal models. *IEEE Visualization*, pp. 357–362.

Hoppe H 1996 Progressive meshes. *ACM SIGGRAPH*, pp. 99–108.

Hoppe H and Praun E 2005 Shape compression using spherical geometry images. In *Advances in Multiresolution for Geometric Modelling* (ed. Dodgson N, Floater M and Sabin M), pp. 27–46. Springer-Verlag.

Hoppe H, DeRose T, Duchamp T, Halstead M, Jin H, McDonald J, Schweitzer JE and Stuetzle W 1994 Piecewise smooth surface reconstruction. *ACM SIGGRAPH*, pp. 295–302.

Hoschek J, Schneider FJ and Wassum P 1989 Optimal approximate conversion of spline surfaces. *Computer Aided Geometric Design* **6**(4), 293–306.

Huffman DA 1952 A method for the construction of minimum-redundancy codes. Proceedings of the *Institute of Radio Engineer*, pp. 1098–1101.

Ihm I and Park S 1998 Wavelet-based 3D compression scheme for very large volume data. *Graphics Interface*, pp. 107–116.

Isenburg M 2000 Triangle Fixer: edge-based connectivity compression. *European Workshop on Computational Geometry*, pp. 18–23.

Isenburg M 2002 Compressing polygon mesh connectivity with degree duality prediction. *Graphics Interface*, pp. 161–170.

Isenburg M and Alliez P 2002 Compressing polygon mesh geometry with parallelogram prediction. *IEEE Visualization*, pp. 141–146.

Isenburg M and Gumhold S 2003 Out-of-core compression for gigantic polygon meshes. *ACM SIGGRAPH*, pp. 935–942.

Isenburg M and Lindstrom P 2005 Streaming meshes. *IEEE Visualization*, p. 30.

Isenburg M and Snoeyink J 2000 Face Fixer: compressing polygon meshes with properties. *ACM SIGGRAPH*, pp. 263–270.

Isenburg M, Lindstrom P, Gumhold S and Snoeyink J 2003 Large mesh simplification using processing sequences. *IEEE Visualization*, pp. 465–472.

Isenburg M, Lindstrom P and Snoeyink J 2005 Streaming compression of triangle meshes. *Eurographics Symposium on Geometry Processing*, pp. 111–118.

Jeong WK and Kim CH 2002 Direct reconstruction of displaced subdivision surface from unorganized points. *Journal of Graphical Models* **64**(2), 78–93.

Kanai T 2001 Meshtoss – converting subdivision surfaces from dense meshes. *6th International Workshop on Vision, Modeling and Visualization*, pp. 325–332.

Kanai T Ohtake Y and Kase K 2006 Hierarchical error-driven approximation of implicit surfaces from polygonal meshes. *Eurographics Symposium on Geometry Processing*, pp. 21–30.

Karni Z and Gotsman C 2000 Spectral compression of mesh geometry. *ACM SIGGRAPH*, pp. 279–286.

Kazinnik R and Elber G 1997 Orthogonal decomposition of non-uniform B-spline space using wavelets. *Computer Graphics Forum* **16**(3), 27–38.

Keren D and Gotsman C 1999 Fitting curves and surfaces with constrained implicit polynomials. *IEEE Transactions on Pattern Analysis and Machine Intelligence* **21**(1), 31–41.

Keren D, Cooper DB and Subrahmonia J 1994 Describing complicated objects by implicit polynomials. *IEEE Transactions on Pattern Analysis and Machine Intelligence* **16**(1), 38–53.

Khodakovsky A and Guskov I 2003 Compression of normal meshes. *Geometric Modeling for Scientific Visualization*. Springer-Verlag.

Khodakovsky A, Schroder P and Sweldens W 2000 Progressive geometry compression. *ACM SIGGRAPH*, pp. 271–278.

Khodakovsky A, Alliez P, Desbrun M and Schroder P 2002 Near-optimal connectivity encoding of 2-manifold polygon meshes. *Journal of Graphical Models* **64**(3–4), 147–168.

Klein R, Liebich G and Strasser W 1996 Mesh reduction with error control. *IEEE Visualization*, pp. 311–318.

Kobbelt L, Vorsatz J, Labik U and Seidel HP 1999 A shrink wrapping approach to remeshing polygonal surfaces. *Computer Graphics Forum* **18**(3), 119–130.

Krishnamurthy V and Levoy M 1996 Fitting smooth surfaces to dense polygon meshes. *ACM SIGGRAPH*, pp. 313–324.

Kronrod B and Gotsman C 2001 Efficient coding of non-triangular mesh connectivity. *Journal of Graphical Models* **63**, 263–275.

Kronrod B and Gotsman C 2002 Optimized compression of triangle mesh geometry using prediction trees. *International Symposium on 3D Data Processing, Visualization and Transmission*, pp. 602–608.

Lavoué G, Dupont F and Baskurt A 2007 A framework for quad/triangle subdivision surface fitting: application to mechanical objects. *Computer Graphics Forum* **26**(1), 1–14.

Lee A, Sweldens W, Schroder P, Cowsar L and Dobkin D 1998 MAPS: multiresolution adaptive parameterization of surfaces. *ACM SIGGRAPH*, pp. 95–104.

Lee A, Moreton H and Hoppe H 2000 Displaced subdivision surfaces. *ACM SIGGRAPH*, pp. 85–94.

Lee E and Ko H 2000 Vertex data compression for triangular meshes. *IEEE Pacific Graphics*, pp. 225–234.

Lee H, Alliez P and Desbrun M 2002 Angle-Analyzer: a triangle-quad mesh codec. *Computer Graphics Forum* **21**(3), 383–392.

Lei Z and Cooper DB 1998 Linear programming fitting of implicit polynomials. *IEEE Transactions on Pattern Analysis and Machine Intelligence* **20**(2), 212–217.

Leonardis A, Jaklic A and Solina F 1997 Superquadrics for segmenting and modeling range data. *IEEE Transactions on Pattern Analysis and Machine Intelligence* **19**(11), 1289–1295.

Levoy M, Pulli K, Curless B, Rusinkiewicz S, Koller D, Pereira L, Ginzton M, Anderson S, Davis J, Ginsberg J, Shade J and Fulk D 2000 The digital Michelangelo project: 3D scanning of large statues. *ACM SIGGRAPH*, pp. 131–144.

Levy B, Petitjean S, Ray N and Maillot J 2002 Least squares conformal maps for automatic texture atlas generation. *ACM SIGGRAPH*, pp. 362–371.

Li Q, Wills D, Phillips R, Viant WJ, Griffiths JG and Ward J 2004 Implicit fitting using radial basis functions with ellipsoid constraint. *Computer Graphics Forum* **23**(1), 55–69.

Li WC, Ray N and Lévy B 2006 Automatic and interactive mesh to T-spline conversion. *Eurographics Symposium on Geometry Processing*.

Lindstrom P 2000 Out-of-core simplification of large polygonal models. *ACM SIGGRAPH*, pp. 259–262.

Lindstrom P and Turk G 1998 Fast and memory efficient polygonal simplification. *IEEE Visualization*, pp. 279–286.

Litke N, Levin A and Schroder P 2001 Fitting subdivision surfaces. *IEEE Visualization*, pp. 319–324.

Lloyd S 1982 Least square quantization in PCM. *IEEE Transaction as Information Theory* **28**, 129–137.

Loop C 1987 Smooth subdivision surfaces based on triangles. Master's thesis, Utah University.

Lounsbery M, DeRose TD and Warren J 1997 Multiresolution analysis for surfaces of arbitrary topological type. *ACM Transactions on Graphics* **16**(1), 34–73.

Lyche T and Morken K 1992 Spline-wavelets of minimal support. In *Numerical Methods in Approximation Theory* (ed. Braess D and Schumaker L), Birchäuser, pp. 177–194.

Ma W and Kruth J 1995 Parametrization of randomly measured points for the least squares fitting of B-spline curves and surfaces. *Computer-Aided Design* **27**, 663–675.

Ma W, Ma X, Tso SK and Pan Z 2004 A direct approach for subdivision surface fitting from a dense triangle mesh. *Computer Aided Design.* **36**(16), 525–536.

Mallat SG 1989 A theory for multiresolution signal decomposition: the wavelet representation. *IEEE Transactions on Pattern Analysis and Machine Intelligence* **11**(7), 674–693.

Marinov M and Kobbelt L 2005 Optimization methods for scattered data approximation with subdivision surfaces. *Journal of Graphical Models* **67**(5), 452–473.

Martin G 1995 Multiresolution analysis of curves and surfaces for use in terrain representation. PhD thesis, University of Central Florida.

Masuda H, Ohbuchi R and Aono M 1999 Frequency domain compression and progressive transmission of parametric surfaces. *Journal of IPSJ* pp. 1188–1195.

Mignot A and Garneau P 2004 MPEG-4 toward solid representation. *IEEE Transaction as Circuits and Systems for Video Technology* **14**(7), 967–974.

Milroy MJ, Bradley C, Vickers GW and Weir DJ 1995 G1 continuity of B-spline surface patches in reverse engineering. *Computer-Aided Design* **27**(6), 471–478.

Mongkolnam P, Razdan A and Farin G 2003 Lossy 3D mesh compression using loop scheme. *International Conference on Computers, Graphics, and Imaging.*

Morse BS, Yoo TS, Rheingans P, Chen DT and Subramanian KR 2001 Interpolating implicit surfaces from scattered surface data using compactly supported radial basis functions. *Shape Modeling Conference*, pp. 89–98.

MPEG-4 2002a ISO/IEC 14496-11. Scene description and application engine.

MPEG-4 2002b ISO/IEC 14496-16. Coding of audio-visual objects: animation framework extension (AFX).

Muraki S 1991 Volumetric shape description of range data using *Blobby Model. Computer Graphics* **25**(4), 227–235.

Muraki S 1993 Volume data and wavelet transforms. *IEEE Computer Graphics and Applications* **13**(4), 50–56.

Ning P and Hesselink L 1992 Vector quantization for volume rendering. *Symposium on Volume Visualization*, pp. 69–74.

Pajarola R and Rossignac J 2000 Compressed progressive meshes. *IEEE Visualization and Computer Graphics* **6**(1), 79–93.

Pentland A 1990 Automatic extraction of deformable part models. *International Journal of Computer Vision* **4**(2), 107–126.

Peters J 1994 Constructing C^1 surface of arbitrary topology using biquadratic and bicubic splines. In *Designing Fair Curves and Surface*, (ed. Sapidis N) pp. 277–293.

Peters J 1995 Biquartic C^1-surface splines over irregular meshes. *Computer-Aided Design* **27**(12), 895–903.

Peters J 2000 Patching Catmull-Clark meshes. *ACM SIGGRAPH*, pp. 255–258.

Peyré G and Cohen L 2005 Geodesic computations for fast and accurate surface remeshing and parameterization. *Progress in Nonlinear Differential Equations and Their Applications* **63**, 157–171.

Peyré G and Mallat S 2005 Surface compression with geometric bandelets. *ACM SIGGRAPH*, pp. 601–608.

Popovic J and Hoppe H 1997 Progressive simplicial complexes. *ACM SIGGRAPH*, pp. 217–224.

Pottmann H and Leopoldseder S 2003 A concept for parametric surface fitting which avoids the parametrization problem. *Computer Aided Geometric Design* **20**(6), 343–362.

Praun E and Hoppe H 2003 Spherical parametrization and remeshing. *ACM SIGGRAPH*, pp. 340–349.

Ray N, Li WC, Lévy B, Sheffer A and Alliez P 2006 Periodic global parameterization. *ACM Transactions on Graphics* **25**(4), 1460–1485.

Rogers D and Fog N 1989 Constrained B-spline curve and surface fitting. *Computer Aided Geometric Design* **21**, 641–648.

Rossignac J 1999 Edgebreaker: connectivity compression for triangle meshes. *IEEE Transactions on Visualization and Computer Graphics* **5**(1), 47–61.

Sahin T and Unel M 2005 Fitting globally stabilized algebraic surfaces to range data. *IEEE International Conference on Computer Vision*, pp. 1083–1088.

Salomie IA, Munteanu A, Gavrilescu A, Lafruit G, Schelkens P, Deklerck R and Cornelis J 2004 MESHGRID – a compact, multiscalable and animation-friendly surface representation. *IEEE Transactions on Circuits and Systems for Video Technology* **14**(7), 950–966.

Samozino M, Alexa M, Alliez P and Yvinec M 2006 Reconstruction with Voronoi centered radial basis functions. *Eurographics Symposium on Geometry Processing*, pp. 51–60.

Sander PV, Snyder J, Gortler SJ and Hoppe H 2001 Texture mapping progressive meshes. *ACM SIGGRAPH*, pp. 409–416.

Santa-Cruz D and Ebrahimi T 2002 Compression of parametric surfaces for efficient 3D model coding. *SPIE Visual Communications and Image Processing*, pp. 280–291.

Sarkar B and Menq CH 1991 Parameter optimization in approximating curves and surfaces to measurement data. *Computer Aided Geometric Design* **8**(4), 267–290.

Savchenko V, Pasko A, Okunev O and Kunii T 1995 Function representation of solids reconstructed from scattered surface points and contours. *Computer Graphics Forum* **14**(4), 181–188.

Schroeder W, Zarge J and Lorensen W 1992 Decimation of triangle meshes. *ACM SIGGRAPH*, pp. 65–70.

Sederberg TW, Zheng J, Bakenov A and Nasri A 2003 T-splines and T-NURCCS. *ACM SIGGRAPH*, pp. 477–484.

Shannon CE 1948 A mathematical theory of communication. *Bell Systems Technical Journal* **27**, 379–423, 623–656.

Solina F and Bajcsy R 1990 Recovery of parametric models from range images: the case for superquadrics with global deformations. *IEEE Transactions on Pattern Analysis and Machine Intelligence* **12**(2), 131–147.

Sorkine O, Cohen-Or D and Toldeo S 2003 High-pass quantization for mesh encoding. *Eurographics Symposium on Geometry Processing*, pp. 42–51.

Susuki H 1999 Subdivision surface fitting to a range of points. *IEEE Pacific Graphics*, pp. 158–167.

Szymczak A, King D and Rossignac J 2001 An Edgebreaker-based efficient compression scheme for regular meshes. *Computational Geometry* **20**(1–2), 53–68.

Szymczak A, Rossignac J and King D 2003 Piecewise regular meshes: construction and compression. *Journal of Graphical Models* **64**(3–4), 183–198.

Taubin G 1995 A signal processing approach to fair surface design. *ACM SIGGRAPH*, pp. 351–358.

Taubin G and Rossignac J 1998 Geometric compression through topological surgery. *ACM Transactions on Graphics* **17**(2), 84–115.

Taubin G, Cukierman F, Sullivan S, Ponce J and Kriegman DJ 1994 Parameterized families of polynomials for bounded algebraic curve and surface fitting. *IEEE Transactions on Pattern Analysis and Machine Intelligence* **16**, 287–303.

Taubin G, Guéziec A, Horn W and Lazarus F 1998 Progressive forest split compression. *ACM SIGGRAPH*, pp. 123–132.

Terzopoulos D and Metaxas D 1991 Dynamic 3D models with local and global deformations: deformable superquadrics. *IEEE Transactions on Pattern Analysis and Machine Intelligence* **13**(7), 703–714.

Tong Y, Alliez P, Cohen-Steiner D and Desbrun M 2006 Designing quadrangulations with discrete harmonic forms. *Eurographics Symposium on Geometry Processing*.

Touma C and Gotsman C 1998 Triangle mesh compression. *Graphics Interface*, pp. 26–34.

Turan G 1984 Succinct representations of graphs. *Discrete Applied Mathematics* **8**, 289–294.

Turk G 1992 Re-tiling polygonal surfaces. *ACM SIGGRAPH*, pp. 55–64.

Turk G and O'Brien JF 1999 Shape transformation using variational implicit functions. *ACM SIGGRAPH*, pp. 335–342.

Valette S and Chassery JM 2004 Approximated centroidal Voronoi diagrams for uniform polygonal mesh coarsening. *Computer Graphics Forum* **23**(3), 381–389.

Valette S and Prost R 2004a Wavelet based multiresolution analysis of irregular surface meshes. *IEEE Transactions on Visualization and Computer Graphics* **10**(2), 113–122.

Valette S and Prost R 2004b A wavelet-based progressive compression scheme for triangle meshes: Wavemesh. *IEEE Transactions on Visualization and Computer Graphics* **10**(2), 123–129.

Witten IH, Neal RM and Cleary JG 1987 Arithmetic coding for data compression. *Communications of the ACM* **30**(6), 520–540.

Wu J and Kobbelt L 2003 A stream algorithm for the decimation of massive meshes. *Graphics Interface*, pp. 185–192.

Yeo B and Liu B 1995 Volume rendering of DCT-based compressed 3D scalar data. *IEEE Transactions on Visualization and Computer Graphics* **1**(1), 29–43.

Yngve G and Turk G 2002 Robust creation of implicit surfaces from polygonal meshes. *IEEE Transactions on Visualization and Computer Graphics* **8**(4), 346–359.

3

3D Indexing and Retrieval

Stefano Berretti, Mohamed Daoudi, Alberto Del Bimbo, Tarik Filali
Ansary, Pietro Pala, Julien Tierny, Jean-Phillippe Vandeborre

3.1 Introduction

This chapter introduces the problem of 3D retrieval by shape descriptors
and the criteria needed to evaluate the 3D retrieval algorithms. The role
of shape descriptors is to represent the original data in a very short way.
Intuitively, this means that the index should be invariant to some geometric
transformations of the object (translation, rotation, scaling), and should have
a certain robustness to noise.

This chapter also details different criteria commonly used to compare 3D
shape descriptors. We distinguish two main criteria: *performance criteria*
that evaluate quantitatively and objectively the performances of the 3D shape
descriptors, and *algorithmic criteria* that evaluate the algorithmic properties
of the 3D shape descriptors.

The use of 3D image and model databases throughout the Internet is growing in both number and size. Indeed, the development of modeling tools,
3D scanners (Figure 3.1), 3D graphic accelerated hardware, Web3D, high-
quality PDAs (Personal Data Assistants) and even cellular phones with a fast
CPU are powerful enough to visualize 3D models interactively. Recently,

3D Object Processing: Compression, Indexing and Watermarking   J.-L. Dugelay, A. Baskurt & M. Daoudi
© 2008 John Wiley & Sons, Ltd

Figure 3.1 A 3D scanner

Acrobat 3D software proposed to insert and to publish 3D designs from major computer-aided design (CAD) applications in Adobe PDF documents.

Exploiting the information contents of digital collections poses several problems. In order to create value out of these collections, users need to find information that matches certain expectations, a notoriously hard problem due to the inherent difficulties of describing visual information content. Nowadays, people facing the problem of finding multimedia information are typically using text-based search engines. The content of a multimedia database is described using every day or – more or less – technical words depending on the usage of the database. Hence, this kind of search engine relies on human operators, with a certain expertise in the domain concerning the multimedia database, who are manually describing the multimedia content using keywords or annotations. The end-user is also experimenting with the problem of language level depending on his or her technical skills and expertise in the domain. Moreover, textual descriptions of multimedia content are inherently subjective and consequently are not a reliable solution to the problem of multimedia data indexing and retrieval.

In recent years, as a solution to these problems, many systems have been proposed that enable effective information retrieval from digital collections of images and videos (Del Bimbo 1999; Stamou and Kollias 2005).

Figure 3.2 Example of a similarity search on a database of 3D models, showing a query object and a set of relevant retrieval results

However, solutions proposed so far to support retrieval of images and videos are not always effective in application contexts where the information is intrinsically 3D. A similarity metric has to be defined to compute the visual similarity between two 3D models, given their descriptions.

Figure 3.2 illustrates the concept of content-based 3D retrieval. The query object is a 3D model of an aircraft (left side of Figure 3.2). The system is expected to retrieve similar 3D models from the database as shown on the right side of Figure 3.2.

Recent advances in 3D scanner acquisition technology and 3D graphics rendering have boosted the creation of 3D model archives for several application contexts. These include:

- **Games/entertainment**: The 3D models are used to enhance realism in entertainment applications. Reuse and adaptation of 3D models by similarity search in existing databases is a promising approach to reduce production costs.
- **Medicine**: In medicine, the detection of similar 3D organ deformation can be used for diagnostics.
- **CAD**: Technicians and engineers in manufacturing companies need to exploit the large CAD mechanical parts database during the design of an automobile, to update the documentary corpus, to exploit engineering data and to create spare parts from original vehicle parts. Presently, many of

these steps demand human intervention for manual and visual inspection. Consequently, these processes are time consuming and very expensive. To improve productivity it is important to develop CAD search algorithms so as to automate significant parts of the process. Indeed, when a new product is designed, it can be composed of many small parts that fit together to form the product. If some of these parts are similar to one of the standard parts already designed, then the possible replacement of the original part with the standard part can lead to a reduction in production costs.

- **3D-face recognition**: Automatic face recognition has been actively researched in recent years, and various techniques using ideas from 2D image analysis have been presented. Although significant progress has been made, the task of automated, robust face recognition is still a distant goal. The 2D image-based methods are inherently limited by variability in imaging factors such as illumination and pose. An emerging solution is to use laser scanners for capturing the surfaces of human faces, and use these data to perform face recognition. Such observations are relatively invariant to illumination and pose, although they do vary with facial expressions. As the technology for measuring facial surfaces becomes simpler and cheaper, the use of 3D facial scans will be increasingly prominent.

- **Cultural heritage**: One particular interest is the possibility of exploiting 3D-based technologies to enable monitoring, cataloguing and remote fruition (periodic acquisition and comparison to evidence for deformation of masterpieces due to pollution, inadequate preservation, etc.).

- **Bioinformatics**: A 3D structural comparison and structural database searching of proteins play important roles. In many cases, merely comparing the amino acid sequences of the proteins cannot provide sufficient information required by the biologist. In particular, one cannot detect the similarity of two remotely homologous proteins by sequence comparison alone. Instead, a comparison of their 3D structures needs to be made in order to determine their similarity since the 3D structures are better preserved than the sequences throughout evolution. Indeed, there are now more than 25 000 protein structure files in Protein Data Bank[1] (PDB), with an additional 100 added every week – hence the increasing necessity for protein structural retrieval.

A variety of retrieval methods have been proposed that enable the efficient querying of model repositories for a desired 3D shape, many of which use a 3D model as a query and attempt to retrieve models with matching shape

---

[1] http://www.pdb.org/.

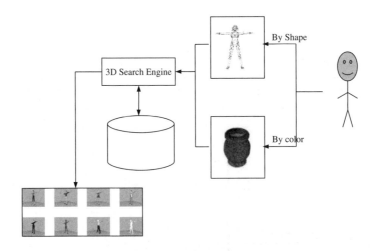

Figure 3.3   Examples of visual query

from the database. As above, an example of such an application is shown in Figure 3.2. The user specifies an aircraft as a query model (left). The system then compares the query to every model in the database, returning pointers to the models that are most similar (right).

In practice, the visual descriptor (shape, color) (Figure 3.3) is incorporated into the search engine as shown in Figure 3.4. In this chapter, we will restrict our discussion to an analysis of shapes to study their contribution in indexing. In a pre processing step the shape descriptor of each model in the database is computed (*offline step*). Then, at run-time, a query is presented to the system, its shape descriptor is computed (*online step*), the query descriptor is compared to the descriptors of the models in the database (3D search engine), and the database models with descriptors that are most similar to the query descriptor are returned as matches.

Figure 3.4 shows an overview of the 3D retrieval system:

- A subsystem for the manual annotation of the 3D data; in general, all the concepts which cannot be extracted manually.
- A subsystem for the extraction of low-level features; 3D processing is used to extract automatically descriptors of 3D data.
- An interface for browsing the 3D model database.
- A graphical query interface which permits retrieval of 3D models by a drawing or sketch, by a photo or by a 3D model. Figure 3.5 shows an example of the 3D retrieval graphical interface.

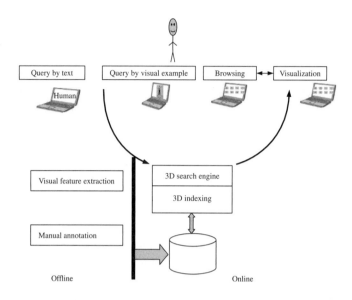

Figure 3.4   An overview of a 3D retrieval system

Figure 3.5   Examples from the Princeton Search Engine of 3D retrieval by sketch (2D and 3D sketch)

3.1.1   3D Shape-invariant Descriptors and Algorithmic Comparison Criteria of 3D Descriptors

Given 3D models, the goal now is to develop metrics and mechanisms for comparing their shapes.

Definition 3.1.1 *The shape is all the geometrical information that remains when location, scale and rotational effects are filtered out from an object (Kendall 1984).*

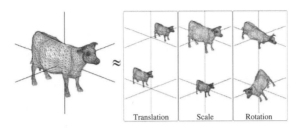

Figure 3.6 Some 3D models which have different locations, rotations and scales but the same shape. Courtesy of M. Kazhdan

So, an object's shape is invariant under the Euclidean similarity transformations of translation, scaling and rotation. This is reflected in Figure 3.6.

Definition 3.1.2 *A sequence of scalars* $\{I_k\}$ *is an invariant descriptor set with regard to a group of transformations G if and only if, for two 3D models* O_1 *and* O_2 *having the same shape,* $I_k(O_1) = I_k(O_2)$ *for all integer k.*

Two objects have the same shape if they can be translated, scaled and rotated to each other so they match exactly, i.e. if the objects are similar. In practice, we are interested in comparing objects with different shapes and so we require a way of measuring shape, some notion of distance between two shapes. This distance should be relatively easy to compute. The 3D scans of objects provide large amounts of data (approximately 20–30 000 vertices, edges, and so on), and a successful method should be able to analyze these data efficiently and compute the metric quickly.

It is important that the invariant descriptors fulfill certain criteria such as real-time computation, completeness and stability which express the fact that a small distortion of the shape does not induce a noticeable divergence. The descriptors must be independent of similarity transformations.

Several invariant descriptors have been proposed in the literature (Bustos *et al.* 2005). To compare the invariant descriptors, six algorithmic criteria have been defined: complexity, invariance under transformations, conciseness, robustness to noise as shown in Figure 3.7, etc, as follows:

- Size, T, representing the size of the vector corresponding to the descriptor of the 3D model.
- Extraction complexity, CE, representing the complexity of the algorithm that extracts the 3D model descriptor.

Figure 3.7 A 3D model from the Princeton Shape Benchmark with decimation and different noise effects

- Comparison complexity, *CMS*, representing the complexity of the similarity measure between two descriptors.
- Generality, *G*, specifying if the 3D model descriptor can be applied to topologically ill-defined 3D models and polygon soup.
- Geometrical invariance, *GI*, specifying if the descriptor is invariant to geometrical transformations (as illustrated in Figure 3.6).
- Topological invariance, *TI*, specifying if the descriptor is independent of the polygonal representations.

3.1.2 Evaluation Criteria

The efficiency of a 3D retrieval algorithm can be evaluated by the following statistics:

- **Nearest neighbor**: The percentage of the closest matches that belong to the same class as the query. This statistic provides an indication of how well a nearest neighbor classifier would perform. Obviously, an ideal score is 100 %; higher scores are associated with good results.
- **First tier and second tier**: The percentage of models in the query's class that appear within the top K matches, where K depends on the size of the query's class. Specifically, for a class with $|C|$ members, $K = |C| - 1$ for the first tier and $K = 2 \times (|C| - 1)$ for the second tier. The first-tier statistic indicates the recall for the smallest K that could possibly include 100 % of the models in the query class, while the second tier is a little less stringent (i.e. K is twice as big). These statistics are similar to the Bull Eye Percentage Score $K = 2 \times |C|$, which has been adopted by the MPEG-7 visual shape descriptors (SDs). In all cases, an ideal matching result gives a score of 100 %; again higher values indicate better matches.
- *E*-**measure**: A composite measure of the precision and recall for a fixed number of retrieved results. The intuition is that a user of a search engine is more interested in the first page of query results than in later pages. So, this measure considers only the first 32 retrieved models for every query

and calculates the precision and recall over those results. The E-Measure is defined as:

$$E = \frac{2}{\frac{1}{P} + \frac{1}{R}}$$

The E-measure is equivalent to subtracting van Rijsbergen's definition of the E-measure from 1. The maximum score is 1.0, and higher values indicate better results.

- **Discounted cumulative gain (DCG)**: A statistic that weights correct results near the front of the list more than correct results later in the ranked list under the assumption that a user is less likely to consider elements near the end of the list. Specifically, the ranked list R is converted to a list G, where element G_i has a value of 1 if element R_i is in the correct class and 0 otherwise. Discounted cumulative gain is then defined as follows:

$$\text{DCG}_1 = G_1; \quad \text{DCG}_i = \text{DCG}_{i-1} + \frac{G_i}{\log_2(i)}, \quad \text{if } i > 1$$

The result is then divided by the maximum possible DCG (i.e. that would be achieved if the first C elements were in the correct class, where $|C|$ is the size of the class) to give the final score:

$$\text{DCG} = \frac{\text{DCG}_k}{1 + \sum_{j=2}^{|C|} \frac{1}{\log_2(j)}}$$

where k is the number of models in the database. The entire query result list is incorporated in an intuitive manner by the discounted cumulative gain (Leifman *et al.* 2003).

- **Recall vs. precision curves**: These curves are well known in the literature of content-based search and retrieval. The recall and precision are defined as follows:

$$\text{Recall} = N/Q, \quad \text{Precision} = N/A$$

where N is the number of relevant models retrieved in the top A retrievals, and Q is the number of relevant models in the collection, which is the number of models to which the query belongs.

Exercise 1 *Figure 3.8 shows five models of the ant–insect class from the* Princeton Shape Benchmark *classified by the user. The user provides the request illustrated in Figure 3.8 (top left) and the 3D search engine gives the first 16 results represented by Figure 3.9. Compute the statistics for nearest*

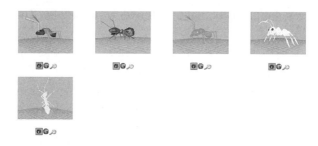

Figure 3.8   Insect class

Figure 3.9   The first 16 relevant retrieval results

neighbor, first tier, second tier, E-measure, and draw the recall/precision curve.

3.2   Statistical Shape Retrieval

The 3D shape retrieval methods are not directly based on some measurements of shapes, but on the distribution of those measurements. This kind of distribution is generally represented by histograms of local or global features. The features could be curvatures on the surface representing the model, the Euclidean or geodesic distance between points on the surface, angle measurements, elementary volumes, and so on. Each 3D model is then described as a distribution of such features. Therefore the principle of comparing two 3D shapes is very simple because it is reduced to a simple comparison of histograms.

In this section, we describe and analyze some methods for computing statistical 3D shape descriptors and dissimilarity measures. The first part introduces some differential geometry definitions. The second and third parts of this section are respectively about local and global features used to compute a histogram. The fourth part discusses the use of what we call hybrid approaches that consist of mixing histograms of local and global features to enhance the possibilities of these two kinds of features. The last part reports on the experiments and results of these three families of methods on a collection of 50 3D models.

3.2.1   Differential Geometries of Surfaces

Let S be a surface denoting a 3D model. Although in practice S is a triangulated mesh with a collection of edges and vertices, we start the discussion by assuming that it is a continuous surface. We recall the main surface properties.

Definition 3.2.1 *A subset S of $\mathbb{R}^3$ is called a regular surface if for each point p in S, there exists a neighborhood V of p in $\mathbb{R}^3$ and a map $\phi : U \to \mathbb{R}^3$ of an open set U, subset $\mathbb{R}^2$, onto V intersection S such that*

1. *ϕ is differentiable.*
2. *$\phi : U \to V \cap S$ is a homeomorphism.*
3. *Each map $\phi : U \to S$ is a regular patch.*

Any open subset of a regular surface is also a regular surface.

Property 1 *If $f : U \to \mathbb{R}$ is a differentiable function in an open set U of $\mathbb{R}^2$, then the graph of f, i.e. the subset of $\mathbb{R}^3$ given by $(x, y, f(x, y))$ for $(x, y) \in U$, is a regular surface.*

Definition 3.2.2 *Let $S \subset R^3$, be a regular and orientable surface. The principal curvatures k^1 and k^2 are the eigenvalues of the Weingarten endomorphism W defined by:*

$$W = I^{-1} \cdot II$$

where I and II are the two fundamental forms (do Carmo 1976) defined as follows:

$$I = \begin{bmatrix} \langle S_u, S_u \rangle & \langle S_u, S_v \rangle \\ \langle S_u, S_v \rangle & \langle S_v, S_v \rangle \end{bmatrix}, \qquad II = \begin{bmatrix} \langle S_{uu}, N \rangle & \langle S_{uv}, N \rangle \\ \langle S_{uv}, N \rangle & \langle S_{vv}, N \rangle \end{bmatrix}$$

where the scalar $\langle a, b \rangle$ represents the scalar product of two vectors a and b. The vector $a \wedge b$ represents the cross-product of two vectors a and b:

$$S_u = \frac{\partial S(u_0, v_0)}{\partial u}, \quad S_v = \frac{\partial S(u_0, v_0)}{\partial v}, \quad N = \frac{S_u \wedge S_v}{\|S_u \wedge S_v\|},$$

$$S_{uu} = \frac{\partial^2 S}{\partial u^2}, \quad S_{vv} = \frac{\partial^2 S}{\partial v^2} \quad \text{and} \quad S_{uv} = \frac{\partial^2 S}{\partial u v}$$

Exercise 2 *The surface of the cylinder (see Figure 3.12) admits the parametrization:*

$$S(u, v) = (\cos u, \sin u, v), U = \left\{ (u, v) \in \mathbb{R}^3; 0 < u < 2\pi, -\infty < v < \infty \right\}$$

Compute the values $S_u, S_{uu}, S_{uv}, S_{vv}, S_u, S_{uu}, S_{uv}, S_{vv}$.

3.2.2 Local Approaches

The local feature histograms aim at using local feature measurements on the surface representing the 3D shape. These features can be understood as information from the 3D model when it is observed and analyzed very closely.

Many local features have been studied in the literature for use as local descriptors for distribution: elementary volumes extracted from each face of a 3D mesh (Zhang and Chen 2001), model area as a function of spherical angles (Ankerst *et al.* 1999) or curvatures extracted at each vertex of a 3D mesh (Zaharia and Prêteux 2001) are some well-known examples. It is also important to note that the curvature histogram has been chosen as an MPEG-7 3D descriptor and has been called the *3D shape spectrum descriptor* in this context.

Shape index (Koenderink and van Doorn 1992)

The 3D shape spectrum descriptor aims at providing an intrinsic shape description of 3D mesh models. The histogram is based on the *shape index* introduced by (Koenderink and van Doorn 1992). The shape index is defined

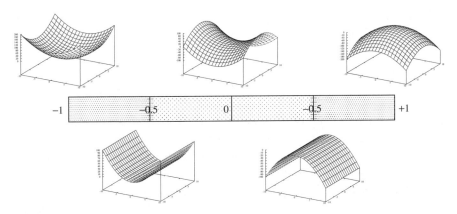

Figure 3.10   The curvature index values for some well-known curves. From left to right: spherical cup, rut, saddle, ridge, spherical cap

as a function of the two principal curvatures of the surface. The main advantage of this index is that it gives the possibility to describe the shape of the object at a given point. The drawback is that it loses the information about the magnitude of the surface shape.

Let us give more details about the computation of the curvature histogram. Let p be a point on the 3D surface. Let us denote by k_p^1 and k_p^2 the principal curvatures associated with the point p. The shape index at point p, denoted by I_p, is defined as:

$$I_p = \frac{2}{\pi} \arctan \frac{k_p^1 + k_p^2}{k_p^1 - k_p^2} \qquad \text{with} \quad k_p^1 \geq k_p^2 \qquad (3.1)$$

The shape index value belongs to the interval $[0, 1]$ and is not defined for planar surfaces. Special values, outside of the interval, can be chosen for special surfaces like planes. Figure 3.10 illustrates some well-known surfaces and their values over the interval of shape index definition.

The shape spectrum of the 3D mesh is the histogram of the shape indexes calculated over the entire 3D mesh.

Estimation of the principal curvatures

Very often a surface is given as the graph of a differentiable function. Let $z = f(x, y)$ belong to an open set $U \subset \mathbb{R}^2$, and $S(u, v) = (u, v, f(u, v))$, $(u, v) \in U$, where $x = u, v = y$. A simple computation shows that:

$$S_u = (1, 0, f_u), \quad S_u = (0, 1, f_v), \quad S_{uu} = (0, 0, f_{uu}), \quad S_{uv} = (0, 0, f_{uv}),$$
$$S_{vv} = (0, 0, f_{vv})$$

$$I = \begin{bmatrix} 1 + f_u^2 & f_u f_v \\ f_u f_v & 1 + f_v^2 \end{bmatrix}, \quad II = \frac{1}{\sqrt{(1 + f_u^2 + f_v^2)}} \begin{bmatrix} f_{uu} & f_{uv} \\ f_{uv} & f_{vv} \end{bmatrix}$$

The estimation of the principal curvatures at a point p is the key step of the shape spectrum extraction. Indeed, the spectrum extraction performances strongly depend on the accuracy of the estimation. Computing these curvatures can be done in different ways, each with advantages and drawbacks. Stokely and Wu (1992) proposed five practical methods to compute them: the Sander–Zucker approach (Sander and Zucker 1990), two methods based on direct surface mapping, a piece wise linear manifold technique, and a turtle geometry method. All these methods are well adapted to specialized uses (e.g. tomographic medical images).

Here, we choose to present a very simple and efficient method to compute the curvature at each vertex of the 3D mesh by fitting a quadric to the neighborhood of this vertex using the least squares method.

The parametric surface approximation is achieved by fitting a quadric surface through the cloud of the m points $\{(x_i, y_i, z_i)\}_{i=1}^m$ made at the centroids of the considered face and its 1-adjacent faces. Here, f is expressed as a second-order polynomial:

$$f(x, y) = a_0 x^2 + a_1 y^2 + a_2 xy + a_3 x + a_4 y + a_5$$

where the a_i coefficients are real values.

By denoting $a = (a_0, a_1, a_2, a_3, a_4, a_5)^t$ and $b(x, y) = (x^2, y^2, xy, x, y, 1)^t$, the previous equation can be expressed by using the standard matrix notation:

$$f(x, y) = a^t \cdot b(x, y)$$

The parameter vector $a = (a_0, a_1, a_2, a_3, a_4, a_5)^t$ is determined by applying a linear regression procedure. Given the data points denoted by $\{(x_i, y_i, z_i)\}_{i=1}^m$, the parameter vector corresponding to the optimal fit (in the mean squared error sense) is given by the following equation:

$$a = \left(\sum_{i=1}^N b(x_i, y_i) b^t(x_i, y_i) \right)^{-1} \cdot \left(\sum_{i=1}^N z_i b(x_i, y_i) \right)$$

From this quadric approximation, it becomes easy to compute the principal curvatures k^1 and k^2.

The curvature histogram

The shape index can now be computed by Equation (3.1) at each point of the 3D mesh. The different I_p values are put together in a histogram representing the curvature distribution for the 3D model.

For more details on curves, surfaces and curvature, we refer the reader to the reference book on differential geometry of curves and surfaces by do Carmo (1976). Another very interesting reference book has been written by Koenderink (1990) who introduced the shape index detailed in this part.

Invariance and robustness of the curvature histogram

Because the shape index (I_p value) used to compute the curvature histogram is a local descriptor, it uses only the neighborhood of the point to be calculated. This neighborhood is obviously invariant to rigid translations and rotations.

The I_p value is calculated with a quotient of the principal curvatures. Then the amplitude of a curve has no influence on the I_p value. Hence, the curvature histogram is also invariant to scaling.

However, the curvature histogram is not robust to tessellation of the 3D mesh. The computation of the shape index depends on the neighborhood of a point. If the resolution of the mesh (i.e. the size of the polygonal faces) is low, then the neighborhood will not be precise enough to approximate the local curve accurately. One idea to fix this is to compute a new tessellation of the 3D mesh with smaller and homothetic triangles.

We summarize the curvature histogram in Algorithm 1.

Algorithm 1   The curvature histogram

Given a surface S:

1. *Compute the values $a_0, a_1, a_2, a_3, a_4, a_5$ of the quadric f on each point of the surface.*

2. *Compute $S_u = (1, 0, f_u)$, $S_u = (0, 1, f_v)$, $S_{uu} = (0, 0, f_{uu})$, $S_{uv} = (0, 0, f_{uv})$, $S_{vv} = (0, 0, f_{vv})$.*

3. *Compute the two fundamental forms I and II, and the matrix W.*

4. *Compute the eigenvalues of the matrix W, and the values of the curvatures k_1 and k_2.*

5. *Compute I_p.*

6. *Compute the curvature histogram.*

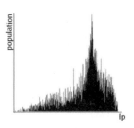

Figure 3.11   Curvature histogram for a 3D model cow (5804 faces)

Figure 3.11 shows an example of a curvature histogram for a 3D-model made of 5804 faces, computed by the algorithm 1.

Exercise 3 *Compute for the cylinder shown in Figure 3.12:*

1. *The first fundamental form.*
2. *The second fundamental form.*
3. *The curvatures k_1 and k_2.*
4. *The shape index I_p.*

The right cylinder over the circle $x^2 + y^2 = 1$ admits the parameterization $\phi : U \to \mathbb{R}^3$, where:

$$\phi(u, v) = (\cos u; \sin u, v),$$

$$U = \left\{ (u, v) \in \mathbb{R}^3; 0 < u < 2\pi, -\infty < v < \infty \right\}$$

Let $R_{x,\theta} : \mathbb{R}^3 \to \mathbb{R}^3$ be the rotation of angle θ about the x axis. Compute the new value of the shape index I_p of the cylinder.

The correlogram of curvature

The main limitation of using curvature histograms for the description and retrieval of 3D models relates to the fact that any information about the spatial distribution of curvature values on the object surface is lost. Correlograms of curvature (Antini *et al.* 2005) have been proposed as a solution to this problem. In particular, with respect to a description based on curvature histograms, correlograms also enable encoding of information about the relative localization of curvature values.

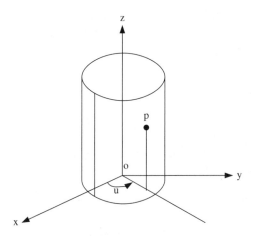

Figure 3.12 Cylinder parameterization

To compute correlograms of curvature, values of the mean curvature are quantized into $2N + 1$ classes of discrete values. For this purpose, a quantization module processes the mean curvature value through a stair–step function so that many neighboring values are mapped to one output value:

$$
\mathcal{Q}(\bar{k}) = \begin{cases} N\Delta & \text{if } \bar{k} > N\Delta \\ i\Delta & \text{if } \bar{k} \in [i\Delta, (i+1)\Delta) \\ -i\Delta & \text{if } \bar{k} \in [-i\Delta, -(i+1)\Delta) \\ -N\Delta & \text{if } \bar{k} < -N\Delta \end{cases} \tag{3.2}
$$

with $i \in \{0, \ldots, N-1\}$ and Δ a suitable quantization parameter (in the experiments reported in the following, $N = 100$ and $\Delta = 0.15$). Function $\mathcal{Q}(\cdot)$ quantizes values of $\bar{k}$ into $2N + 1$ distinct classes $\{c_i\}_{i=-N}^{N}$.

To simplify the notation, $v \in S_i$ is synonymous with $v \in S$ and $\mathcal{Q}(\bar{k}_v) = c_i$.

The correlogram of curvature is defined with respect to a predefined distance value δ. In particular, the correlogram of curvature $\gamma_{c_i c_j}^{(\delta)}$ of a mesh S is defined as:

$$
\gamma_{c_i, c_j}^{(\delta)}(S) = \Pr_{v_1, v_2 \in S}[(v_1 \in S_{c_i}, v_2 \in S_{c_j}) \mid \|v_1 - v_2\| = \delta]
$$

In this way, $\gamma_{c_i, c_j}^{(\delta)}(S)$ is the probability that two vertices that are δ far away from each other have curvature belonging to class c_i and c_j, respectively.

Ideally, $||v_1 - v_2||$ should be the geodesic distance between vertices v_1 and v_2. However, this can be approximated by the k-ring distance if the mesh S is regular and triangulated.

Definition 3.2.3 (1-ring) *Given a generic vertex $v_i \in S$, the neighborhood or 1-ring of v_i is the set:*

$$V^{v_i} = \{v_j \in S : \exists e_{ij} \in E\}$$

where E is the set of all mesh edges (if $e_{ij} \in E$ there is an edge that links vertices v_i and v_j).

The set V^{v_i} can be easily computed using the morphological operator *dilate*:

$$V^{v_i} = \text{dilate}(v_i)$$

Practically, the dilation operator is obtained by using a 1-ring as 'structure element'. A set of mesh vertices is dilated by extending the set with the 1-ring of any vertex of the set (Rossl *et al.* 2000).

Through the dilate operator, the concept of 1-ring can be used to define, recursively, a generic kth-order neighborhood:

$$\text{ring}_k = \text{dilate}^k \cap \text{dilate}^{k-1}$$

The definition of the kth-order neighborhood enables the definition of a true metric between vertices of a mesh. This metric can be used for the purpose of computing curvature correlograms as an approximation of the usual geodesic distance (which is computationally much more time consuming). According to this, we define the k-ring distance between two mesh vertices as $d_{ring}(v_1, v_2) = k$ if $v_2 \in \text{ring}_k(v_1)$.

Function $d_{ring}(v_1, v_2) = k$ is a true metric. In fact:

1. $d_{ring}(u, v) \geq 0$ and $d_{ring}(u, v) = 0$ if and only if $u = v$.
2. $d_{ring}(u, v) = d_{ring}(v, u)$.
3. $\forall w \in S, \quad d(u, v) \leq d(u, w) + d(w, v)$.

Based on the $d_{ring}(\cdot)$ distance, the correlogram of curvature can be redefined as follows:

$$\gamma_{c_i, c_j}^{(k)}(S) = \Pr_{v_1, v_2 \in S}\left[(v_1 \in S_{c_i}, v_2 \in S_{c_j}) \,\big|\, d_{ring}(v_1, v_2) = k\right]$$

Figure 3.13 Correlograms of curvature extracted from sample 3D models: (a) statue; (b) dinosaur

Figure 3.13 shows some 3D models belonging to two classes of objects: statue and dinosaur. For each object the corresponding correlogram of curvature is also shown. It can be noted that values of the correlograms extracted from objects of the same class exhibit a common pattern.

Several distance measures have been proposed to compute the dissimilarity of distribution functions. In order to compute the similarity between curvature correlograms of two distinct meshes $\gamma_{c_i,c_j}^{(k)}(S_1)$ and $\gamma_{c_i,c_j}^{(k)}(S_2)$, we experimented with the following distance measures:

- *Minkowski-form distance*:

$$d_{\mathcal{L}_p} = \left[\sum_{i,j=-N}^{N} \left| \gamma_{c_i,c_j}^{(k)}(S_1) - \gamma_{c_i,c_j}^{(k)}(S_2) \right|^p \right]^{1/p}$$

- *Histogram intersection*:

$$d_{HI} = 1 - \frac{\sum_{i,j=-N}^{N} \min \left(\gamma_{c_i,c_j}^{(k)}(S_1), \gamma_{c_i,c_j}^{(k)}(S_2) \right)}{\sum_{i,j=-N}^{N} \gamma_{c_i,c_j}^{(k)}(S_2)}$$

- χ^2-*statistics*:

$$d_{\chi^2} = \sum_{i,j=-N}^{N} \frac{\left(\gamma_{c_i,c_j}^{(k)}(S_1) - \gamma_{c_i,c_j}^{(k)}(S_2) \right)^2}{2 \left(\gamma_{c_i,c_j}^{(k)}(S_1) + \gamma_{c_i,c_j}^{(k)}(S_2) \right)}$$

- *Kullback–Leibler divergence*:

$$d_{KL} = \sum_{i,j=-N}^{N} \gamma_{c_i,c_j}^{(k)}(S_1) \log \frac{\gamma_{c_i,c_j}^{(k)}(S_1)}{\gamma_{c_i,c_j}^{(k)}(S_2)}$$

3.2.3   Global Approaches

Global descriptors are a means to handle the global nature of the object. This means that, rather than the details of the object, more importance is given to its general aspect. The moments (Lo and Don 1989; Sadjadi and Hall 1980) here are the traditional mathematical tool. However, the moments present a drawback: they rely on the object being described by a function, defined in major 3D space, which could for instance associate 1 or 0 to each 3D point, depending on whether it is inside or outside the object. Very few objects are defined by or convertible to such a function, which makes the moments method often impossible to use. To avoid this problem, it is also possible to grab the Z-buffer of an object's view, which consists of the 2D data that could be used to generate the moments. But in this section we want to deal with 3D space, so this method is not presented here.

Other global descriptor distributions have been studied in the literature. The most interesting ones have been proposed and developed by Osada *et al.* (2001). They are based on shape distributions. The main idea is to focus on the statistical distributions of a shape function measuring geometrical properties of the 3D model. They are represented as histograms, as in the local approach. The range of possible functions is very wide. The authors have proposed and tested five shape functions:

- A3: measures the angle between three random points on the surface of a 3D model;
- D1: measures the distance between a fixed point (e.g. the centroid of the boundary of the model) and one random point on the surface;
- D2: measures the distance between two random points on the surface;
- D3: measures the square root of the area of the triangle formed by three randomly chosen points on the surface;
- D4: measures the cube root of the volume of the tetrahedron formed by four randomly chosen points on the surface.

These shape functions have been chosen for their computational simplicity and invariance (see below for a discussion on the invariance and robustness of the distance histogram). According to Osada *et al.* (2001), the distribution of the distance between pairs of random points (D2) gives the best results compared to other simple methods.

Other authors have investigated the D2 descriptor to enhance its already very good performance. Ip *et al.* (2002) refined the D2 descriptor in the context of CAD models with a classification of the two random points used to compute the descriptor. Ohbuchi *et al.* (2003a) also extended the D2 descriptor with a distribution called the Absolute Angle–Distance Histogram. This histogram is parameterized by one parameter denoting the distance between two random points (as in the original D2 shape descriptor) and by another parameter denoting the angle between the surfaces on which the two random points are located.

In the next sections, we give more details about the distance histogram (D2), its computation and its properties.

Distance histogram (D2)

The computation of the distance histogram is based on a stochastic method and is particularly simple.

Two random faces of the 3D model are taken, then two random points are taken on those two faces. Finally, the Euclidean distance between those two points is computed. The method is iterated N times, N being big enough to give a good approximation of the distribution. Empirically, Osada *et al.* (2001) found that $N = 1024^2$ is a good compromise between the computation time and storage and the precision of the distribution.

Figure 3.14 shows an example of a curvature histogram for a 3D model made of 5804 faces. Figure 3.15 shows distance histograms for some canonical shapes.

Invariance and robustness of the distance histogram (D2)

It can be seen that all the shape functions proposed by Osada *et al.* (2001) – including D2 – are invariant to rigid motions (translations and rotations). They are also robust to tessellation of the 3D mesh, since points are randomly selected on the surface of the 3D model. They are also relatively insensitive to small perturbations due to noise, cracks, etc., since sampling is area weighted.

But one should notice that the Euclidean distance, and thus the D2 index, are sensitive to scaling. This is obviously because the Euclidean distance is not either. So, the histogram has to be normalized before using it. One way to do this is to normalize the mean value of the distribution.

This value is first computed:

$$\mathrm{MV}(f) = \int_0^\infty x \cdot f(x) \cdot dx$$

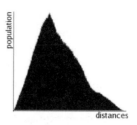

Figure 3.14 Distance histogram (D2) for the 3D model of a cow (5804 faces), O. Croquette, J-P. Vandeborre, M. Daoudi, C. Chaillou "Indexing and retrieval VRML models"
SPIE Electronic Imaging 2002, in proceedings volume 4672, pp. 95–106. Reproduced by
permission of SPIE

Then the new histogram is defined by:

$$f_{norm}(x) = f\left(\frac{x}{\mathrm{MV}(f)}\right)$$

The f_{norm} is the new distribution, invariant to scaling.

3.2.4 Hybrid Approaches

The previously presented local and global methods have advantages and
drawbacks (see the conclusion of this section). Some authors proposed
hybrid – or *mixing* – approaches to enhance advantages and to reduce the
problems induced by these methods.

Paquet *et al.* (2000) proposed to use different descriptors: namely, bounding boxes, cords based, moments based and wavelet based. Paquet and Rioux
(1997) also proposed, in the context of the Nefertiti project, to combine geometry and color/texture appearance to describe a 3D-model and express a user
request.

More in keeping with the previously presented local and global approaches,
Vandeborre *et al.* (2002) described the 3D models with three distributions:
a curvature histogram, a distance histogram and an elementary volume histogram. In order to compare two 3D models, the authors compute three
distances between pairs of similar histograms (one distance for the curvature histograms, one distance for the distance histograms and one distance for the elementary volume histograms) as shown in Figure 3.16. Then,
the rank of each object is calculated according to each descriptor, sorting them by decreasing values (three integers between 1 and *NbObjects*,

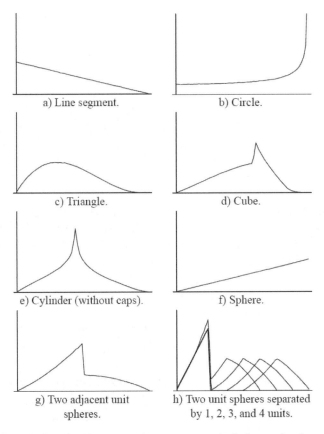

a) Line segment.

b) Circle.

c) Triangle.

d) Cube.

e) Cylinder (without caps).

f) Sphere.

g) Two adjacent unit
spheres.

h) Two unit spheres separated
by 1, 2, 3, and 4 units.

Figure 3.15 D2 distance histograms for some canonical shapes Osada *et al*. (2001).
© 2001 IEEE

named $Rank_c$ for the curvature index, $Rank_d$ for the distance index and
$Rank_v$ for the volume index). Those values are merged into a single value,
using a formula. Intuitively, there are different strategies for merging these
values:

- Support the objects having satisfactory results with one approach, the other
  ones having less importance. This is what the authors called the 'OR'
  method:

$$F = 1 - \left(\frac{(Rank_s - 1) \cdot (Rank_d - 1) \cdot (Rank_v - 1)}{NbObjects \times NbObjects \times NbObjects} \right)$$

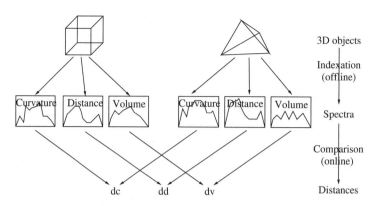

Figure 3.16 Comparison of two 3D models described by three histogram. O. Croquette, J-P. Vandeborre, M. Daoudi, C. Chaillou "Indexing and retrieval VRML models" *SPIE Electronic Imaging 2002, in proceedings volume 4672, pp. 95–106.* Reproduced by permission of SPIE

- Use the mean of the three results, called the 'MEAN' method:

$$F = \frac{(Rank_s + Rank_d + Rank_v)}{3 \cdot NbObjects}$$

This way, they obtain one final real number, between 0 and 1, which represents the confidence one could have in the result. Note that the same kind of formula can be used to merge just two rank values.

3.2.5 Experiments and Results

The methods described in the previous sections (i.e. curvature histogram, distance histogram, mixing methods) have been implemented and tested on a 3D model collection containing 50 models. Experiments and results are briefly reported in this section to show the effectiveness, advantages and drawbacks of the three families of approaches presented in the previous sections.

Before detailing the 3D model test collection and the results, let us explain the way the histograms are compared.

Comparison of histograms

There are several ways to compare distribution histograms: the Minkowski L_n norms, Kolmogorov–Smirnov distance, match distances, and many others.

One of the most used is the L_1 norm because of its simplicity and its accurate results.

The Minkowski L_n norm is given by:

$$D_{L_1}(f_1, f_2) = \int_{-\infty}^{+\infty} |f_1 - f_2|$$

where f_1 and f_2 are the two functions to be compared.

The first step before calculating the Minkowski L_n norm on two histograms is to interpolate them in a fixed number of linear segments by the least squares method in order to circumvent some problems like quantization, noise, etc. Then, a simple integration of these interpolations is done, giving a real number as a result.

3D model test collection

The 3D model test collection contains 50 models which have been collected from the MPEG-7 3D shape core experiments (Zaharia and Prêteux 2000) and arranged into seven classes. This manual classification has only been done for the purposes of our experiment. This classification is never used to improve the retrieval task of the search engine. The classes are the following: A-class (eight *aircraft* objects), M-class (considered just as noise, five *misc.* objects), P-class (seven *chess piece* objects), H-class (eight *human* objects), F-class (six *fish* objects), Q-class (eight *quadrupede* objects) and C-class (eight *car* objects). The models are 3D meshes ('soup' of polygons) of approximately 500 to 25 000 faces. Moreover, the mesh levels of detail are very different from one object to another.

Classification matrix

As the 3D model collection is classified, one can assume, for each request, that objects from the same class as the request are relevant, and consequently others are not. To evaluate the global performances of the different stand-alone descriptors (curvature and distance histogram) and of the mixing methods, classification matrices have been used.

Each column of a classification matrix (e.g. Figure 3.17) represents an object of the database (the first column corresponds to the first object, and so on) which has been used as a request in the search engine. Each small square shows how the object of the given row was ranked: the darker the square, the better the rank. For example, object 23 was ranked last when object 50 was

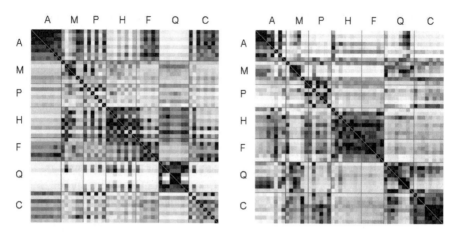

Figure 3.17   Classification matrices for the curvature (left) and distance (right) Histograms. O. Croquette, J-P. Vandeborre, M. Daoudi, C. Chaillou "Indexing and retrieval VRML models" *SPIE Electronic Imaging 2002, in proceedings volume 4672, pp. 95–106.* Reproduced by permission of SPIE

requested, 50 the square at the intersection of column 50 and line 23 is white in color. This also means that the diagonal is entirely black because each object is always the best result of its own request. Moreover, the matrices are not symmetric because the squares do not represent the distance between two objects, but a rank.

Results

The classification matrices for the curvature and distance histograms are shown in Figure 3.17. The first observation is that the two descriptors have difficulties in retrieving some classes. For example, the curvature histogram has no difficulty in classify the aircraft, but has many problems with the car class. Similarly, the distance index gives an accurate classification for the cars, but cannot give satisfactory results for the aircraft. Both curvature and distance indices have advantages and drawbacks.

Firstly, one can see that the two mixing methods are quite equivalent in terms of results. Figure 3.18 shows the classification matrices for the OR and the MEAN methods with the curvature and the distance histograms. Comparing these matrices (Figure 3.18) to the ones for the stand-alone histograms (Figure 3.17), it is clear that the two mixing methods can correct some weaknesses of the stand-alone histogram methods. For example, the curvature

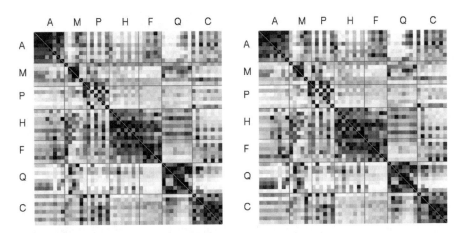

Figure 3.18 Classification matrices for the curvature (left) and distance (right) Histograms. O. Croquette, J-P. Vandeborre, M. Daoudi, C. Chaillou "Indexing and retrieval VRML models" *SPIE Electronic Imaging 2002, in proceedings volume 4672, pp. 95–106.* Reproduced by permission of SPIE

histogram matrix shows some difficulties with the cars, and the distance matrix shows some difficulties with the aircraft. Those problems no longer exist when any of the mixing methods are used. On the other hand, the curvature histogram is able to classify humans and fishes, whereas the distance histogram has some problems with these classes. So, the mixing methods cannot entirely organize these two classes correctly. Nevertheless, any mixing methods give better results than any of the single histogram methods.

3.2.6   Conclusion on Statistical Methods

Statistical methods have the main advantage of reducing the 3D model similarity measure to a very simple comparison of histograms. The previous shape functions used to compute these histograms generally have the long-awaited properties of invariance to rotation and to translation. Invariance to scaling is generally a matter of histogram normalization. Local approaches (such as the curvature histogram) are able to distinguish different classes of objects, but are sensitive to noise. Global approaches are robust to noise and can distinguish models in wide categories, but are not efficient at discriminating objects that are globally similar but with different small details in their shapes. Hybrid methods can be used to combine different statistical descriptors and enhance the performance of 3D model retrieval.

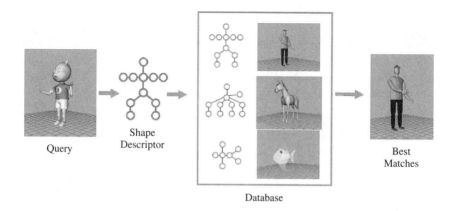

Figure 3.19 The method of 3D retrieval parts Courtesy of M Kazhdan

3.3 Structural Approaches

The aim of this section is to present structural methods for 3D retrieval. These methods use high-level information to describe the structure of 3D objects. We will mainly present the graph-based methods, as shown in Figure 3.19 illustrating a 3D retrieval system based on such information. The statistical shape descriptors represent the geometry of the model and they can be used to compare rigid body transformations and to compare whole models. However, they cannot be used to compare articulated models and parts of a model. The general approach proposes to construct graphs where nodes represent parts and edges represent relationships between parts. The proposed algorithms compare nodes and match graphs.

In this section, we are interested in the topology of the shape. Indeed, the topology describes properties of shape that are invariant under articulations. We can investigate topology by investigating critical points of Morse functions.

Definition 3.3.1 (Morse function) *Let $p(u) \in M \subset \mathbb{R}^3$ be a point on a closed embedded 2-manifold, in a neighborhood continuously parameterized by $u = (u_1, u_2)$. Let $f : M \to \mathbb{R}$ be any smooth function on the manifold. A point is critical if its gradient $\partial f / \partial u$ vanishes; otherwise, it is regular. A critical point is Morse if its Hessian matrix $H(p)$ is non-singular; otherwise, it is degenerate. If and only if all its critical points are Morse, then the function f is a Morse function.*

The Hessian of the function f at the point p is the matrix of second derivatives defined by:

$$
\begin{bmatrix}
\frac{\partial^2 f}{\partial u_1 \partial u_1}(p) & \frac{\partial^2 f}{\partial u_1 \partial u_2}(p) & \cdots & \frac{\partial^2 f}{\partial u_1 \partial u_d}(p) \\
\frac{\partial^2 f}{\partial u_2 \partial u_1}(p) & \frac{\partial^2 f}{\partial u_2 \partial u_2}(p) & \cdots & \frac{\partial^2 f}{\partial u_2 \partial u_d}(p) \\
\vdots & & \ddots & \\
\frac{\partial^2 f}{\partial u_d \partial u_1}(p) & \frac{\partial^2 f}{\partial u_d \partial u_2}(p) & \cdots & \frac{\partial^2 f}{\partial u_d \partial u_d}(p)
\end{bmatrix}
$$

Exercise 4 *Compute the Hessian and, if defined, the index of the origin, which is critical for each function in the list below:*

1. $f(u_1, u_2) = u_1^2 + u_2^2$.
2. $f(u_1, u_2, u_3) = u_1 u_2 + u_1 u_3 + u_2 u_3$.

3.3.1 Introduction to Reeb Graph

A Reeb graph (Reeb 1946) is a topological structure that encodes the connectivity relations of the critical points of a Morse function defined on an input surface.

The Reeb graph is a schematic way of presenting a Morse function where the vertices of the graph are critical points and the arcs of the graph are connected components of the level sets of f, contracted to points. More formally, Reeb graphs are defined as follows.

Definition 3.3.2 (Reeb graph) *Let $f : M \to \mathbb{R}$ be a simple Morse function defined on a compact manifold M. The Reeb graph of f is the quotient space of f in $M \times \mathbb{R}$ from the equivalence relation $(p_1, f(p_1)) \sim (p_2, f(p_2))$, if and only if:*

$$
\begin{cases}
f(p_1) = f(p_2) \\
p_1 \text{ and } p_2 \text{ belong to the same connected} \\
\quad \text{component of } f^{-1}(f(p_1))
\end{cases}
$$

Figure 3.20 gives an example of a Reeb graph computed on a bi-torus with regard to the height function and illustrates well the fact that Reeb graphs can be used as skeletons.

Constructing a Reeb graph from a Morse function f computed on a triangulated surface first requires identification of the set of vertices corresponding to critical points. With this aim, several formulations have been proposed (Edelsbrunner and Mücke 1990; Takahashi *et al.* 1995) to identify local

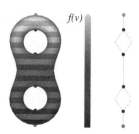

Figure 3.20   Evolution of the level lines, using the height function as $f(v)$, on a bi-torus, its critical points and its Reeb graph, J. Tierny, J-P. Vandeborre, M. Daoudi "Invariant high-level Reeb graphs of 3D polygonal meshes" IEEE 3DPVT 2006, 3rd International Symposium on 3D Data Processing, Visualization and Transmission, Chapel Hill, North Carolina, USA, June 14–16, 2006, pp. 105–112. © 2006 IEEE

maxima, minima and saddles, observing for each vertex the evolution of f at its direct neighbors. Several algorithms have been developed to construct Reeb graphs from the connectivity relations of these critical points (Carr *et al.* 2004; Cole-McLaughlin *et al.* 2003), most of them in $O(n \times log(n))$ steps, with n the number of vertices in the mesh.

However, they assume that all the information brought by the Morse function f is relevant (Carr *et al.* 2004; Ni *et al.* 2004). Consequently, they assume that all the identified critical points are meaningful, while in practice this hypothesis can lead to intractably large Reeb graphs. To overcome this issue, (Ni *et al.* 2004) developed a user-controlled simplification algorithm. (Bremer *et al.* 2004) proposed an interesting critical point cancellation technique based on a *persistence* threshold. (Attene *et al.* 2003) proposed a seducing approach, unifying the graph construction and simplification, but it is conditioned by a *slicing* parameter.

Lemma 1 *The Reeb graph of a Morse function on a compact, closed, orientable 2-manifold of genus g has g loops.*

Exercise 5 *Consider the Morse function $f : S \to \mathbb{R}$ whose level lines have been drawn in Figure 3.21. Compute the Reeb graph for the following surface (Figure 3.21).*

3.3.2   Reeb Graphs for Shape Matching

A Reeb graph encodes the behavior of a Morse function on the shape and also tells us about the topology of the shape. The main properties of 3D shape

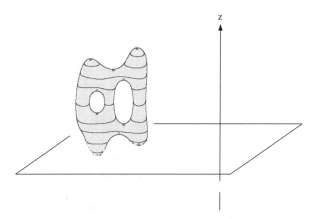

Figure 3.21 Compute the Reeb graph of the Morse function f defined on this surface (exercise 5)

matching algorithms are the invariance to rotation and robustness to noise and variability in model pose. However, the height function is not appropriate for 3D shape matching because it is not invariant to transformations such as rotation. In order to avoid these problems many authors propose to use a geodesic distance: that is, the distance from point to point on a surface. Figure 3.22 gives an example of geodesic distance computed between two

Figure 3.22 Example of geodesic distance

points. Using geodesic distance provides rotation invariance and robustness against problems due to noise.

Hilaga *et al.* (2001) presented an approach to describe the topology of 3D models by a graph structure and show how to use it for matching and retrieval. They proposed the geodesic function μ defined by:

$$\mu(v) = \int_{p \in s} g(v, p) ds$$

where the function $g(v,p)$ returns the distance between v and p on s. The discrete version of this function is defined by:

$$\mu : \mu(v) = \sum g(v, b_i) \times \text{area}(b_i)$$

In Tierny *et al.* (2007b), first the mesh feature points in Figure 3.23(a) are automatically extracted, intersecting geodesic-based map extrema. Then, for each vertex in the mesh, a mapping function f_m is defined as the geodesic distance to the closest feature point (Figure 3.22). Next, for each vertex $v \in S$, an upper value approximation of $f_m^{-1}(f_m(v))$, denoted $\Gamma(v)$, is computed along the edges of S. In particular, the connected component of $\Gamma(v)$ containing v is identified and denoted $\gamma(v)$. Analyzing the evolution of the number of connected subsets of $\Gamma(v)$ as f_m evolves enables the construction of a Reeb graph (Reeb 1946) (Figure 3.23(b)). At this stage, each connected component of the Reeb graph is modeled by an ordered collection of closed curves $\gamma(v)$. The next step consists of identifying *constrictions* (Hétroy 2005) within these collections. For each connected component of the Reeb graph, the average Gaussian curvature on each curve $\gamma(v)$ is computed. Then, local

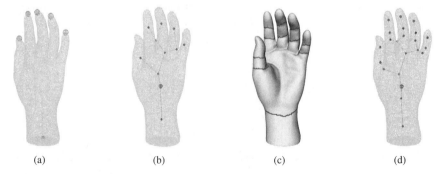

(a)   (b)   (c)   (d)

Figure 3.23   Enhanced topological skeleton extraction Tierny *et al.* (2007b). © 2007 IEEE

(a)   (b)   (c)   (d)

Figure 3.24 Segmentation of a hand surface model into its sub parts Tierny *et al.* (2007a). © 2007 Blackwell Publishing

negative minima are identified as constrictions (Figure 3.23(c)). Finally, the connected components of the Reeb graph are subdivided using these constrictions as boundaries between sub parts (Figure 3.23(d)). As a conclusion of this algorithm, the input surface is represented by an enhanced topological skeleton (Figure 3.23(d)), which encodes the topological and geometrical evolution of the contours of the mapping function f_m.

Once the Reeb graph is computed, the surface is segmented to a set of sub parts of the surface. Figure 3.24 shows an example of the results obtained by this algorithm.

Tierny *et al.* (2007a) proposed a new idea to compare two surfaces. It is based on the key idea that two surface models are similar if their sub parts are similar. For each subpart its signature is computed as a function of the distortion introduced by its mapping to its canonical planar domain.

They proposed to compare two subpart signatures by using an L_1 distance. Then, they computed the distance between two models by running a bipartite matching algorithm (Tam and Lau 2007) that matches pairs of topology-equivalent sub parts that minimize their distance, minimizing the overall sum of distances, denoted d. Finally, the distance between the two models is given by d.

Figure 3.25 shows a typical query and the results retrieved by the system. Sub parts that have been matched together have been displayed with the same gray level. Notice that, except in one case, the tail of the horse query model has been matched by the tail of each retrieved result. Similar comments can be made for the legs, or the neck, which demonstrates the efficiency of the proposed signature. Moreover, this figure shows that the proposed signature is clearly pose insensitive, since horses in different poses have been retrieved as the top results.

Figure 3.26 zooms in on the subpart matchings between a hand and a human surface model, displaying some of the sub parts that have been

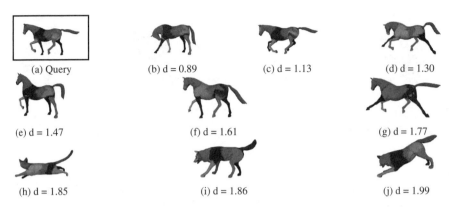

(a) Query   (b) d = 0.89   (c) d = 1.13   (d) d = 1.30

(e) d = 1.47   (f) d = 1.61   (g) d = 1.77

(h) d = 1.85   (i) d = 1.86   (j) d = 1.99

Figure 3.25 Subpart similarity matchings between a horse query model and the retrieved results Tierny *et al*. (2007a). © 2007 Blackwell Publishing

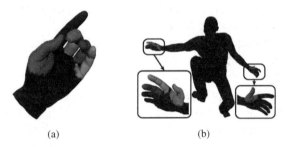

(a)   (b)

Figure 3.26 Subpart similarity matchings between a hand and a human model Tierny *et al*. (2007a). © 2007 Blackwell Publishing

matched together. Notice that the thumb has been matched with the correct thumb of the humanoid and that the remaining fingers have been matched with the fingers of the humanoid, despite their difference in pose. These results can be extended using graph-based matching algorithms so that, for example, the hand model can be matched by its corresponding hand in the humanoid, achieving partial shape similarity (Funkhouser *et al.* 2004).

3.4 Transform-based Approaches: Spin Images

This section provides an overview of transform-based 3D shape descriptors. These descriptors, such as spherical harmonics, spin images, Zernike 3D, etc, are based on the transformation of the 3D shape from 3D Euclidean space representation to another space representation.

A description of 3D models based on spin images has been proposed by Johnson and Hebert (1999) with the aim of supporting a view-independent recognition of objects by capturing both their global and local features.

Such a description of 3D models based on spin images can be regarded as a compromise between view-based and structure-based approaches to object description. On the one hand, a description based on spin images shares relevant common traits with view-based approaches. For instance, each spin image is obtained by projecting information of interest according to a particular viewpoint on the object surface. On the other hand, information that is projected on a spin image does not reflect just object parts that are visible from a given viewpoint as is typical of view-based approaches. Rather, projected data encode geometric features of visible as well as cluttered object parts.

Basically, spin images encode the density of mesh vertices projected onto a 2D space: the 3D mesh vertices are first mapped onto a 2D space defined with respect to one point in the 3D space; then, the resulting coordinates are used to build a 2D histogram.

More precisely, let $O = \langle p, n \rangle$ be an *oriented point* on the surface of the object, where p is a point on the surface of the object and n the normal of the tangent plane in p. For a generic oriented point O, a *spin map* can be defined, which maps any point x in the 3D space onto a 2D space according to the following formula (see also Figure 3.27 for the notation):

$$S_O(x) \rightarrow [\alpha, \beta] = [\sqrt{\|x - p\|^2 - (n \cdot (x - p))^2}, n \cdot (x - p)]$$

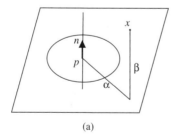 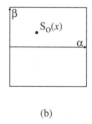

(a)   (b)

Figure 3.27   Given an oriented point $\langle p, n \rangle$ on the object surface, a generic point x is mapped on to point $[\alpha, \beta]$ on the spin map, $[\alpha, \beta]$ being the radial distance and the elevation of x w.r.t. to $\langle p, n \rangle$: (a) the object-centered 3D coordinate system, (b) the spin map coordinate system, "Content-based Retrieval of 3D Models A. Del Bimbo, P. Pala ACM Transactions on Multimedia Computing, Communications and Applications, Vol. 2, No. 1, February 2006", 2006 ACM, Inc. Reprinted by permission

In other words, the oriented point defines a family of cylindrical coordinate systems, with the origin at p, and with the axis along n. The spin map projection of x retains the radial distance (α) and the elevation (β), while it discards the polar angle. In so doing, the ambiguity is resolved, which results from the fact that the oriented point does not define a unique cylindrical coordinate system. This projection ensures that, for any given oriented point, a unique spin map exists.

To produce a *spin image* of an object, a spin map is applied to points of the object surface. Hence, given a mesh representation of the object, the spin image is obtained by applying the spin map to vertices of the mesh. Since several vertices can be mapped into the same (α, β) coordinates, the spin image retains information about the number of vertices mapped onto each (α, β) value. Hence, the spin image is a 2D histogram $I(i, j)$ that is constructed by projecting coordinates α and β of each mesh vertex according to a bilinear interpolation scheme. The purpose of the interpolation scheme is to spread the contribution of each vertex over the nearest points on the grid obtained by quantizing values of α and β. By spreading the contribution of each vertex, a smooth representation is obtained in contrast to the 'dotted' map that would be obtained by simply accumulating indexed values of α and β (see Figure 3.28).

Most relevant characteristics of spin images are invariance to rigid transformations (as a consequence of the adoption of an object-centered coordinate system), limited sensitivity to variations in the position of mesh vertices

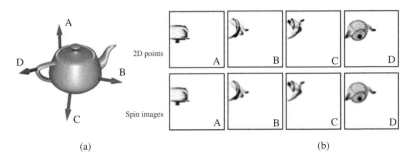

Figure 3.28 A 3D object (a), together with 2D point images and spin images computed from four distinct points on the object surface, 2006 ACM, Inc. Reprinted by permission. "Content-based Retrieval of 3D Models A. Del Bimbo, P. Pala ACM Transactions on Multimedia Computing, Communications and Applications, Vol. 2, No. 1, February 2006", 2006 ACM, Inc. Reprinted by permission

(which might result from the adoption of different sampling schemes), flexibility (since no hypotheses are made on the surface representation) and ease of computation.

Object description based on spin images entails a huge amount of information which makes it difficult to use them for applications addressing retrieval by similarity from large databases rather than object recognition in 3D scenes. To overcome these limitations, spin image signatures have been proposed in Assfalg *et al.* (2007). Spin image signatures make use of feature extraction and clustering techniques to meet the storage and efficiency requirements of content-based retrieval from large 3D model repositories.

3.5   View-based Approach

The human visual system has an uncanny ability to recognize objects from single views, even when presented monocularly under fixed viewing conditions. For example, the identity of most of the 3D models in Figure 3.29 is obvious.

The issue of whether 3D model recognition should rely on internal representations that are inherently 3D or on collections of 2D views has been explored by Riesenhuber and Poggio (2000). They showed that, in a human visual system, a 3D model is represented by a set of 2D views.

As shown in Figure 3.30, the process of 3D model comparison using views can be broken into two phases: a phase of indexing and a phase of retrieval. In the phase of indexing, for every 3D model of the database, we calculate the

Figure 3.29   Some 3Dmodels from the Princeton Shape Benchmark. Reproduced from Princeton Shape Benchmark

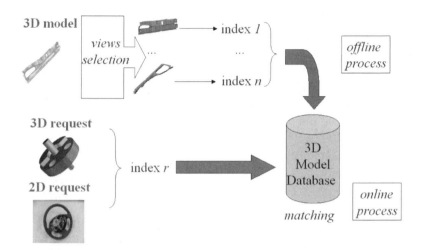

Figure 3.30 The view selection process Filali Ansary *et al.* (2007). © 2007 IEEE

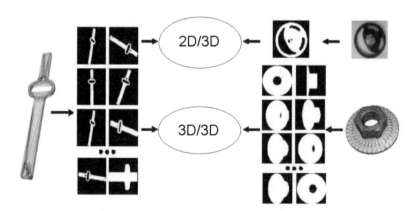

Figure 3.31 A 3D model characterized by a set of 2D views and two types of 3D retrieval. Courtesy of J Ricard

characteristic views and their associated descriptors. In the retrieval phase, the query undergoes a similar treatment as the 3D models of the database, following which a descriptor (invariant to some deformations) is calculated and compared to the descriptors of the database objects. Figure 3.31 shows an example of a 3D model characterized by a set of 2D views and the two

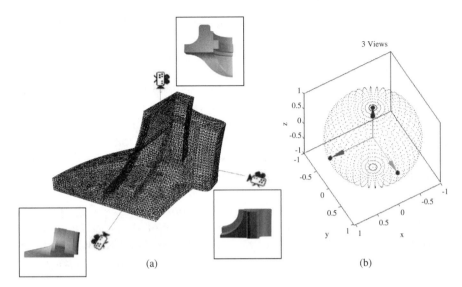

(a)   (b)

Figure 3.32 Three views created around the 3D model corresponding to three principal axes. Courtesy of J Ricard

types of realizable retrieval: 2D/3D by comparing the set of view descriptors to a query image: and 3D/3D by comparing the two sets of views.

The process of comparing two 3D models using their views can be separated in to two main steps:

- **3D model indexing**: Each 3D model is characterized by a set of 2D views. For each view, a 2D descriptor is calculated.
- **3D model retrieval**: The same descriptor is applied to the 2D query views. A matching process is then started to match the request 2D views to the views representing the 3D models of the database.

The main idea of this type of method is to represent a 3D model by a set of characteristic views. For that, the space of views of the 3D model is discretized to N points of view distributed around the 3D model. For each point, a view (2D image) of the 3D model is taken. Figure 3.32 shows three views created around the 3D model by placing the cameras on the three principal axes.

Each view is indexed using a method of 2D image analysis. In the retrieval process, each view of the 3D model request is indexed and compared to the N views of each 3D model. The two main points discussed in the 3D retrieval literature are the selection of the characteristic views and the choice of the descriptor. The choice of the view descriptor will not modify the principle of the view-based approach, but can modify the performance and the effectiveness of it.

The main idea of view-based similarity methods is that two 3D models are similar, if they look similar from all viewpoints. This paradigm leads to the implementation of query interfaces based on defining a query by one or more views, sketches, photos showing the query from different points of view. Figure 3.30 shows an overview of the characteristic view selection algorithm. The main difficulty in view-based 3D retrieval is to characterize a 3D model by a set of 2D views.

One point which will have an impact on the effectiveness and the speed of the view-based methods is the number and position of views used to describe a 3D model. This number is directly related to the performance of the system. Two approaches were used to choose the number of characteristic views:

- Using a fixed and restricted number of views.
- Using a dynamic number of views.

3.5.1   Methods with a Fixed Number of Views

To guarantee a low number of views and consequently quick retrieval, a class of techniques can create a restricted number of views with fixed positions, independently of the 3D model to be indexed. These methods select a fixed number of views on a sphere surrounding the object and create images starting from these points of view. To increase the similarity between views, the 3D models are aligned and scaled. For this, the principal axes of the objects are calculated by analyzing the principal component (PCA), for example, or by the calculation of the matrix of covariances. Figure 3.33 shows four examples of views distributed on the viewing sphere.

Light-field descriptor

Chen *et al.* (2003) proposed a descriptor used for comparing the similarity among 3D models extracted from 4D light fields, which are representations of a 3D model. The light fields describe the radiometric properties of light in space. A light field (or plenoptic function) is traditionally used in image-based rendering and is defined as a 5D function that represents the radiance at a given 3D point in a given direction. For a 3D model, the representation is the same along a ray, so the dimension of the light fields around an object

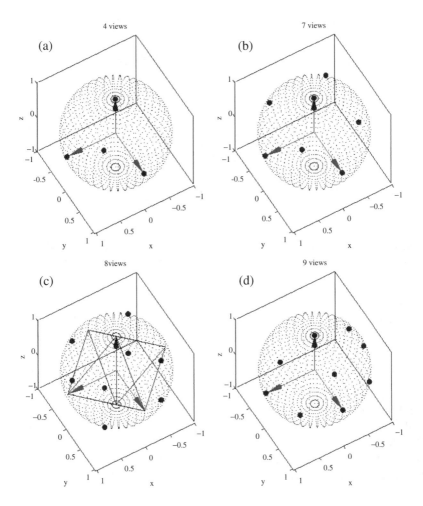

Figure 3.33 Example of views distributed on the viewing sphere. Courtesy of J. Ricard

can be reduced to four. Each 4D light field of a 3D model is represented by a collection of 2D images, which are rendered from a 2D array of cameras. The camera positions of one light field can be put either on a flat surface or on a sphere in the 3D world. The light-field representation has not only been used in image-based rendering, but also in image-driven simplification to decide which portions of an object to simplify.

The main idea comes from the following statement: 'If two 3D models are similar, they also look similar from all viewing angles.' Accordingly,

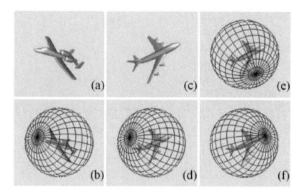

Figure 3.34 The main ideas for measuring similarity between two 3D models using a Light-field descriptor. (D Chen *et al*. 2003 © 2003 Blackwell Publishing)

the similarity between two 3D models can be measured by summing the similarity from all corresponding images of a light field. However, what must be considered is the transformation, including translation, rotation and scaling.

The camera system surrounding each model is rotated until the highest overall similarity (cross-correlation) between the two models from all viewing angles is reached. Take Figure 3.34 as an example, where (a) and (c) are two different aircraft with inconsistent rotations. Firstly, for the aircraft in (a), the cameras of a light field are placed on a sphere, as shown in (b), they are put on the intersection points of the sphere. Then, the cameras of this light field can be applied, at the same positions, to the aircraft in (c), as shown in (d). By summing the similarities of all pairs of corresponding images in (b) and (d), the overall similarity between the two 3D models is obtained. Next, the camera system in (d) can be rotated to a different orientation, such as in (e), which leads to another similarity value between the two models. After evaluating similarity values, the correct corresponding orientation, in which the two models look most similar from all corresponding viewing angles, can be found, such as in (f). The similarity between the two models is defined by summing the similarity from all corresponding images between (b) and (f):

1. Since each model has its own coordinate system, translation and scaling are applied first in order to ensure that a model is entirely contained in each rendered image. The input 3D model is translated from the center of

the model to the origin of the world coordinate system, and then scales the axis of maximum length to be 1. The translation $T = (T_x; T_y; T_z)$ assigns the middle point of the whole model to be the new origin:

$$T_i = \frac{MaxCoor_i + MinCoor_i}{2}, \quad i = x, y, z$$

where $MaxCoor_i$ and $MinCoor_i$ are the maximum and minimum coordinate values of the i axis, respectively. The scaling is isotropic, and normalizes according to the maximum distance from the x, y and z axes of the whole model:

$$S = \frac{1}{\min_{i=x,y,z}(MaxCoor_i + MinCoor_i)}$$

2. Images are rendered from the camera positions of the light fields, which are on the surface of a larger sphere. There are 10 light fields for each 3D model, and the camera positions of each light field are set at the 20 vertices of a regular dodecahedron. The camera at each viewpoint is directed toward the center of the sphere, and the up-vectors of the cameras are placed uniformly. If two 3D models are of different orientations, their proper corresponding viewing images will have different rotational angles. This does not matter, since the image metric of the approach is also robust against rotations.
3. An orthogonal projection is used in order to reduce the size of descriptors. Therefore, in a descriptor of a light field, there are 10 images represented from 20 viewpoints. For a 3D model, 10 descriptors of light fields are created, so there are 100 images that should be rendered and extracted for features.
4. The Zernike moment and Fourier descriptor are extracted from each image. Descriptors for a 3D model are those features from the 100 images.

Depth buffer-based methods

Ohbuchi *et al.* (2003b) proposed a method which works on polygon soup 3D models (MODFD). The 3D models are made invariant to translation and scaling. Then they compute $N = 42$ depth buffer rendered images (a kind of range image) of the model. These sets of images discretely cover all possible view aspects of the model.

Then for each image a Fourier-transform-based descriptor is computed (2D descriptor). The set of these 42 descriptors comprises the multiple oriented shape descriptor for the 3D model. Let M_1 and M_2 be two 3D models to be compared. The similarity measure is defined as follows:

$$D(M_1, M_2) = \frac{1}{N} \sum_{i=1}^{N} \min_{j \in 1 \ldots N} d(M_{1i}, M_{2j})$$

where M_{1i}, M_{2j} are the computed 2D descriptors and the distance d is the L_1 norm. Since there is no ordering to the rotations, the similarity measure compares all possible pairings and picks the one with the minimum L_1 distance to contribute to the sum over all the 42 views.

Vranic (2003) proposed a descriptor based on the *buffer-based descriptor* (BBD) introduced by Heczko *et al.* (2002). To make the descriptor invariant to rotation and scaling, each 3D model is oriented using continuous PCA and normalized for scaling.

Using projections on the bounding box, six depth buffer images are used. Each image is described using a 2D fast Fourrier transform (2D-FFT). The 3D model descriptor is based on the low-frequencies features of the 2D-FFT of each depth buffer image.

Chaouch and Verroust-Blondet (2006) developed a new approach called the *enhanced depth buffer-based descriptor*, based on the depth buffer images and the continuous PCA to be geometrically invariant. This method uses the same descriptor as Vranic (2003), but weights the buffer images used in the descriptor, based on informational criteria.

The approach suggested by Vajramushti *et al.* (2004) also consists of describing the 3D model using depth images. The surface and the volume calculated from these projections constitute the characteristic features vector of the 3D model. These vectors are then used in an iterative algorithm to estimate the errors between the depth images while varying the six degrees of freedom characterizing the 3D model position vector. This approach allows us to find the best corresponding 2D depth images to overcome the PCA problems.

Other methods

We can cite, for example, the work of Abbasi and Mokhtarian (2001), which fixe the number of views associated with each 3D model to nine views: four views corresponding to the top, side, front and back views, and five intermediate views (Figure 3.33(d)). Chen and Stockman (1998) set the number of views to eight views equally distributed on the viewing sphere of the

3D model. The camera is placed on each viewpoint to obtain the views (Figure 3.33(c)).

Nayar *et al.* (1996), as well as Mahmoudi and Daoudi (2007), considered each 3D model as a cloud of points. Principal axes of each 3D model are calculated using the eigenvalues of the matrix of covariance. The 3D model cloud of points is projected in 2D from according to the three principal directions thus calculated. Seven characteristic views are created: three principal views along the principal axes and four views according to the four directions corresponding to 45° views between the principal views (Figure 3.33(b)).

3.5.2 Methods with a Dynamic Number of Views

Conventional multi-view representations are based on a large number of views and cannot be used in many applications such as retrieval from large databases. Multi-view representations have to deal with the following issues:

1. What is the optimal number of views?
2. How to select the optimal views?

To solve these problems some methods have been proposed for automatic selection of optimal views of an object. In order to represent an object efficiently, these methods eliminate similar views and select a relatively small number of views using an optimization algorithm. This number varies depending on the complexity of the object and the measure of expected accuracy.

Aspect graphs

In aspect-graph-based methods, the main idea is that 3D shapes look different when viewed from different viewpoints. For example, a cube looks like a square when viewed from the top. Based on this idea, the space of views can be partitioned into view classes or characteristic views. Within each class, the views share a certain property. A clustering algorithm might be used to generate the view classes.

A view class representation called an aspect graph was proposed by Koenderink (1990). The nodes of the graph represent the aspects, namely a class of views, and the edges connect different nodes which have a certain change in aspect. These appearance changes from node to node are called visual events. Aspect graphs are complicated data structures, therefore their usage is limited.

Using aspect graphs, Cyr and Kimia (2001) specified a query by a view of 3D models. A descriptor of a 3D model consists of a set of views. The number of views is kept small by clustering them and by representing each cluster by one view, which is described by a shock graph. Schiffenbauer (2001) presented a complete survey of aspect graph methods. Using shock matching, Macrini *et al.* (2002) applied indexing using topological signature vectors to implement view-based similarity matching more efficiently.

Adaptive views clustering

In the adaptive views clustering (AVC) presented by Filali Ansary *et al.* (2007), the main idea is to generate an initial set of views from the 3D model, and then reduce this set to only those that best characterize this 3D model. This idea comes from the fact that not all the views of a 3D model are of equal importance: there are some views that contain more information than others. For example, one view is sufficient to represent a sphere because it looks the same from all angles, but more than one view is needed to represent a more complex 3D model such as an aircraft.

In the AVC method, every 2D view is represented by 49 Zernike moment coefficients. To choose X characteristic views which best represent a set of $N = 320$ initial views, the authors presented a derivative of the X-means (Pelleg and Moore 2000) clustering algorithm, where, instead of using a fixed number of clusters, they used the range $[1, \ldots, 40]$, in which they chose the 'optimal' number of clusters.

In essence, the algorithm starts with one characteristic view (K equal to 1), and then adds characteristic views where they are needed. The authors took the global K-means on the data starting with characteristic views as cluster centers. They continued alternating between adding characteristic views and taking global K-means until the upper bound for the characteristic view number (40) was reached. During this process, for each K, they saved the characteristic view set.

To add new characteristic views, they used the idea presented in the X-means clustering method by Pelleg and Moore (2000). Firstly, for every cluster of views represented by a characteristic view, they selected two views that have the maximum distance in this cluster. Next, in each cluster of views, they ran local K-means (with $K = 2$) for each pair of selected views. By *local*, we mean that only the views that are in the cluster are used in this local clustering (Figure 3.35). At this point, a question arises: 'Are the two new characteristic views giving more information on the region than the original

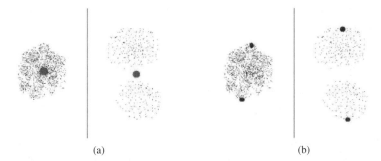

(a)   (b)

Figure 3.35 Local K-means on each part of the views cluster with $K = 2$ Filali Ansary *et al*. (2007), (© 2007 IEEE)

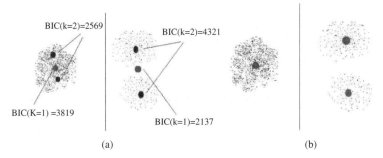

(a)   (b)

Figure 3.36 Selecting the representations (with one or two characteristic views) that have the higher Bayesian information criterion (BIC) score Filali Ansary *et al*. (2007), (© 2007 IEEE)

characteristic view?' To answer this question, the authors used a Bayesian information criterion (BIC) (Schwarz 1978), which scores how likely the representation model (using one or two characteristic views) fits the data. Note that Figures 3.35 and 3.36 are just schematic examples, as we represent a view in a 2D space.

3.6 Normative Aspect: MPEG-7

The MPEG-7 standard, also known as the *Multimedia Content Description Interface,* aims at providing standardized core technologies allowing the description of audiovisual data content in multimedia environments (Salembier and Sikora 2002).

This is a challenging task given the broad spectrum of requirements and targeted multimedia applications, and the broad number of audiovisual features of importance in such contexts. In order to achieve this goal, MPEG-7 will standardize:

- Descriptors (D), representations of features that define the syntax and the semantics of each feature representation.
- Description schemes (DS), schemes that specify the structure and semantics of the relationships between their components, which may be both D and DS.
- A description definition language (DDL), to allow the creation of new DS and, possibly, D and to allow the extension and modification of existing DS.
- System tools, to support multiplexing of description, synchronization issues, transmission mechanisms, file formats, etc.

The description may include still pictures, video, graphics, audio, speech, 3D models and the information about how these elements are put together in a multimedia representation. In the case of 3D data, the 3D shape spectrum descriptor has been proposed as a description of 3D shape. A multiple-views descriptor has also been developed to combine 2D descriptors representing a visual feature of a 3D model seen from different view angles. The descriptor shapes a complete 3D view-based representation of the object. Any 2D visual descriptor, such as contour shape, region shape, color or texture, can be used. The 2D/3D descriptor supports integration of the 2D descriptors used in the image plane to describe features of the 3D (real-world) objects.

3.7 Summary

The use of 3D image and model databases throughout the Internet is growing in both number and size. Exploiting the information content of digital collections poses several problems. In order to create value out of these collections, users need to find information that matches certain expectations, a notoriously hard problem due to the inherent difficulties of describing visual information content.

In recent years, as a solution to these problems, many systems have been proposed that enable effective information retrieval from digital collections of images and videos. However, the solutions proposed so far are not always effective in application contexts where the information is intrinsically 3D.

An indexing algorithm is generally a two-process mechanism:

1. The *off line process* which consists of finding invariant – to certain transformations – and robust – to noise, for example – descriptors for each 3D model in the collection.
2. The *on line process* experimented by the end-user who is querying the search engine to find a relevant 3D model in the collection.

Hence, the two key problems for 3D model indexing are to find performance descriptors and to define a similarity metric in order to compute the visual similarity between 3D models, given their descriptions.

Several approaches can be used to describe a 3D model and were presented in this chapter:

- Statistical approaches focus on the distribution of local and global descriptions of the 3D model surface: local curvatures, global distances, etc.
- Structural approaches aim at finding a high-level structure, such as a graph, from a 3D mesh.
- Transform-based approaches use transformations of the 3D shape from a 3D Euclidean space representation to another space representation.
- View-based approaches defend the intuitive idea that a 3D model is well represented by a set of characteristic views, and use well-known 2D descriptors on these views.

But before presenting these approaches, some comparison and evaluation criteria, so as to compare and evaluate the different 3D indexing methods, have been presented and discussed.

Lastly, the normative aspects were briefly explored for the MPEG-7 standard, also known as the Multimedia Content Description Interface, and how the 3D model domain is integrated in this standard for new multimedia environments.

References

Abbasi S and Mokhtarian F 2001 Affine-similar shape retrieval: application to multi-view 3D object recognition. *IEEE Transactions on Image Processing* **10**, 131–139.
Ankerst M, Kastenmuller G, Kriegel HP and Seidl T 1999 Nearest neighbor classification in 3D protein databases. *7th International Conference on Intelligent Systems for Molecular Biology (ISMB'99)*.

Antini G, Berretti S, Del Bimbo A and Pala P 2005 Retrieval of 3D objects using curvature correlograms. *Proceedings of International Conference on Multimedia & Expo*, Amsterdam.

Assfalg J, Bertini M, Del Bimbo A and Pala P 2007 Content-based retrieval of 3D objects using spin image signatures. *IEEE Transactions on Multimedia* **9**(3), 589–599.

Attene M, Biasotti S and Spagnuolo M 2003 Shape understanding by contour-driven retiling. *The Visual Computer* **19**, 127–138.

Bremer PT, Edelsbrunner H, Hamann B and Pascucci V 2004 Topological hierarchy for functions on triangulated surfaces. *IEEE Transactions on Visualization and Computer Graphics* **10**, 385–396.

Bustos B, Kein DA, Saupe D, Schreck T and Vranić DV 2005 Feature-based similarity search in 3D object databases. *ACM Comput Surv.* **37**(4), 345–387.

Carr H, Snoeyink J and de Panne MV 2004 Simplifying flexible isosurfaces using local geometric measures. *IEEE Visualization*, pp. 497–504.

Chaouch M and Verrous-Blondet A 2006 Enhanced 2D/3D approaches based on relevance index for 3D-shape retrieval. *Shape Modeling International (SMI 2006)*, Matsushima, Japan.

Chen D, Tian X, Shen Y and Ouhyoung M 2003 On visual similarity based 3D model retrieval. *Eurographics*, pp. 223–232, Granada, Spain.

Chen JL and Stockman G 1998 3D free-form object recognition using indexing by contour feature. *Computer Vision and Image Understanding* **71**(3), 334–355.

Cole-McLaughlin K, Edelsbrunner H, Harer J, Natarajan V and Pascucci V 2003 Loops in Reeb graphs of 2-manifolds. *Symposium on Computational Geometry*, pp. 344–350.

Cyr CM and Kimia B 2001 3D object recognition using shape similarity-based aspect graph. *IEEE International Conference on Computer Vision*, pp. 254–261.

Del Bimbo A 1999 *Visual Information Retrieval*. Mogan Kaufmann Publisher, Inc.

do Carmo MP 1976 *Differential Geometry of Curves and Surfaces*. Prentice Hall.

Edelsbrunner H and Mücke EP 1990 Simulation of simplicity: a technique to cope with degenerate cases in geometric algorithms. *ACM Transactions on Graphics* **9**, 66–104.

Filali Ansary T, Daoudi M and Vandeborre JP 2007 A Bayesian 3D search engine using adaptive views clustering. *IEEE Transactions on Multimedia* **9**(1), 78–88.

Funkhouser T, Kazhdan M, Shilane P, Min P, Kiefer W, Tal A, Rusinkiewicz S and Dobkin D 2004 Modeling by example. *ACM Transactions on Graphics* **23**, 652–663.

Heczko M, Keim D, Saupe D and Vranic D 2002 Methods for similarity search on 3D databases. *Datenbank-Spektrum*, pp. 54–63.

Hétroy F 2005 Constriction computation using surface curvature. *Eurographics*, pp. 1–4.

Hilaga M, Shinagawa Y, Kohmura T and Kunii T 2001 Topology matching for fully automatic similarity estimation of 3D shapes. *ACM SIGGRAPH*, pp. 203–212.

Ip CY, Lapadat D, Sieger L and Regli WC 2002 Using shape distributions to compare solid models. *ACM Symposium on Solid Modeling and Applications*, pp. 273–280.

Johnson AE and Hebert M 1999 Using spin-images for efficient multiple model recognition in cluttered 3-D scenes. *IEEE Transactions on Pattern Analysis and Machine Intelligence* **21**(5), 433–449.

Kendall DG 1984 Shape manifolds, procrustean metrics and complex projective spaces. *Bulletin of London Mathematical Society* **16**, 81–121.

Koenderink JJ 1990 *Solid Shape*. The MIT Press.

Koenderink JJ and van Doorn AJ 1992 Surface shape and curvature scales. *Image and Vision Computing* **10**(8), 557–565.

Leifman G, Katz S, Tal A and Mei R 2003 Signatures of 3D models for retrieval. *4th Israel–Korea Bi-National Conference on Geometric Modeling and Computer Graphics*, pp. 159–163.

Lo CH and Don HS 1989 3-D moments forms: Their construction and application to object identification and positioning. *IEEE Transactions on Pattern Analysis and Machine Intelligence* **11**(10), 1053–1064.

Macrini D, Shokoufandeh A, Dickenson S, Siddiqi K and Zucker S 2002 View based 3-D object recognition using shock graphs. *IEEE International Conference on Pattern Recognition* **3**, 24–28.

Mahmoudi S and Daoudi M 2007 A probabilistic approach for 3D shape retrieval by characteristic views. *Pattern Recognition Letters* **28**(13), 1705–1718.

Nayar S, Nene S and Murase H 1996 Real-time object recognition system. *International Conference on Robotics and Automation*.

Ni X, Garland M and Hart J 2004 Fair Morse functions for extracting the topological structure of a surface mesh. *ACM Transactions on Graphics* **23**, 613–622.

Ohbuchi R, Minamitani T and Takei T 2003a Shape similarity search of 3D models by using enhanced shape functions. *Theory and Practice of Computer Graphics*, pp. 97–104.

Ohbuchi R, Nakazawa M and Takei T 2003b Retrieving 3D shapes based on their appearance. *ACM SIGMM Workshop on Multimedia Information Retrieval*, Berkeley, California, pp. 39–46.

Osada R, Funkhouser T, Chazells B and Dobkin D 2001 Matching 3D models with shape distributions. *Shape Modeling International*.

Paquet E and Rioux M 1997 Nefertiti: a query by content software for three-dimensional models databases management. *International Conference on Recent Advances in 3-D Digital Imaging and Modeling*, pp. 345–352.

Paquet E, Murching A, Naveen T, Tabatabai A and Rious M 2000 Description of shape information for 2-D and 3-D objects. *Signal Processing: Image Communication* **16**, 103–122.

Pelleg D and Moore A 2000 X-means: extending k-means with efficient estimation of the number of clusters. *International Conference on Machine Learning*, pp. 727–734.

Reeb G 1946 Sur les points singuliers d'une forme de Pfaff complètement intégrable ou d'une fonction numérique. *Comptes-rendus de l'Académie des Sciences* **222**, 847–849.

Riesenhuber M and Poggio T 2000 Models of object recognition. *Nature Neuroscience* **3**, 50–89.

Rossl C, Kobbelt L and Seidel HP 2000 Extraction of feature lines on triangulated surfaces using morphological operators. *Smart Graphics, Proceedings of the 2000 AAAI Spring Symposium*. Stanford, California.

Sadjadi F and Hall E 1980 Three-dimensional moment invariants. *IEEE Transactions on Pattern Analysis and Machine Intelligence* **2**(2), 127–136.

Salembier P and Sikora T 2002 *Introduction to MPEG-7: Multimedia Content Description Interface*. John Wiley & Sons, Inc.

Sander PT and Zucker SW 1990 Inferring surface trace and differential structure from 3-D images. *IEEE Transactions on Pattern Analysis and Machine Intelligence* **12**(9), 833–854.

Schiffenbauer RD 2001 A survey of aspect graphs. Technical Report TR-CIS-2001-01, CIS.

Schwarz G 1978 Estimating the dimension of a model. *Annals of Statistics* **6**, 461–464.

Stamou G and Kollias S 2005 *Multimedia Content and the Semantic Web*. John Wiley & Sons, Ltd.

Stokely EM and Wu SY 1992 Surface parametrization and curvature measurement of arbitrary 3D objects: five practical methods. *IEEE Transactions on Pattern Analysis and Machine Intelligence* **14**(8), 833–840.

Takahashi S, Ikeda T, Shinagawa Y, Kunii TL and Ueda M 1995 Algorithms for extracting correct critical points and constructing topological graphs from discrete geographical elevation data. *Computer Graphics Forum* **14**, 181–192.

Tam GKL and Lau RWH 2007 Deformable model retrieval based on topological and geometric signatures. *IEEE Transactions on Visualization and Computer Graphics* **13**, 470–482.

Tierny J, Vandeborre JP and Daoudi M 2007a Reeb chart unfolding based 3D shape signatures. *Eurographics*, Prague.

Tierny J, Vandeborre JP and Daoudi M 2007b Topology driven 3D mesh hierarchical segmentation. *IEEE International Conference on Shape Modeling and Applications (SMI'2007)*, Lyons.

Vajramushti N, Kakadiaris I, Theoharis T and Papaioanno G 2004 Efficient 3D retrieval using depth images. *6th ACM SIGMM International Workshop on Multimedia Information Retrieval (MIR'04)*, pp. 189–196.

Vandeborre JP, Couillet V and Daoudi M 2002 A practical approach for 3D model indexing by combining local and global invariants. *IEEE International Symposium on 3D Data Processing, Visualization and Transmission*, pp. 644–647.

Vranic DV 2003 An improvement of rotation invariant 3D shape descriptor based on functions on concentric spheres. *IEEE International Conference on Image Processing*, pp. 757–760.

Zaharia T and Prêteux F 2000 New content for the 3D shape core experiment: the 3D CAF data set. MPEG-7 ISO/IEC JTC1/SC29/WG11 MPEG00/M5915, Noordwijkerhout.

Zaharia T and Prêteux F 2001 3D-shape-based retrieval within the MPEG-7 framework. *SPIE Conference on Nonlinear Image Processing and Pattern Analysis XII*, vol. 4304, pp. 133–145.

Zhang C and Chen T 2001 Efficient feature extraction for 2D/3D objects in mesh representation. *International Conference on Image Processing*.

4

3D Object Watermarking

**Jihane Bennour, Jean-Luc Dugelay, Emmanuel Garcia,
Nikos Nikolaidis**

4.1 Introduction

The digital revolution, the explosion of communication networks and the increasingly growing passion of the general public for new information technologies led to an exponential growth of multimedia document traffic (image, text, audio, video, 3D objects, etc.). This phenomenon is now so important that ensuring protection and control of the exchanged data has become a major issue. Indeed, due to their digital nature, multimedia documents can be duplicated, modified, transformed and shared very easily. In this context, it is important to develop systems for copyright protection and digital rights management (DRM) in general, copy or access control, and content authentication. Traditionally, such issues were handled by data encryption. However, once data were encrypted by an authorized user they could be distributed and manipulated. Watermarking is a promising alternative solution for reinforcing the security of multimedia documents. In this chapter, we will present an overview of the basic notions and principles of 3D object watermarking along with a review of state-of-the-art algorithms.

3D Object Processing: Compression, Indexing and Watermarking   J.-L. Dugelay, A. Baskurt & M. Daoudi
© 2008 John Wiley & Sons, Ltd

4.2 Basic Review of Watermarking

In this section, we recall some basic watermarking notions, using the special case of images, which was historically the first types of digital documents investigated for watermarking (see Figure 4.1).

Image watermarking is a technique that aims at imperceptibly *embedding* secret information in an image according to an optional secret key. It is then possible to check whether secret information has been embedded in the image (i.e. whether the image is watermarked), or whether some specific information (i.e. a message) is actually embedded in the image, or to determine the actual information that was embedded in the image (see Figure 4.2). Thus a watermarking systems consists of two basic modules, namely the watermark embedding (signing) module and the watermark detection/extraction (retrieving) module.

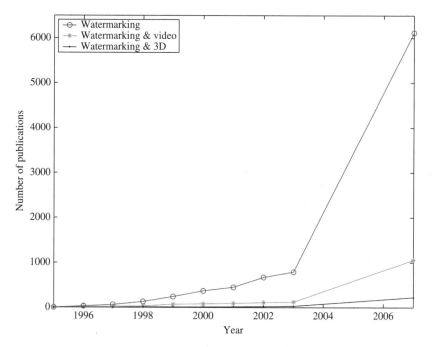

Figure 4.1   Number of watermarking publications registered within the INSPEC database, February 2008. Reproduced from Inspec

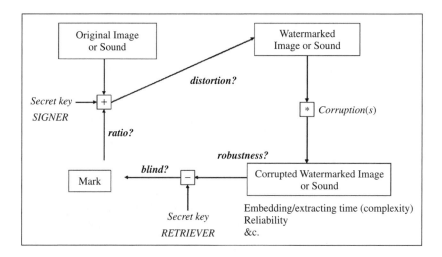

Figure 4.2   Basic scheme of watermarking

Watermarking techniques can be separated into two major categories with respect to the information conveyed by the watermark:

- *Zero bit*: Watermarking techniques in this category can only verify whether the data are watermarked or not.
- *Multiple bit*: These techniques can encode a message consisting of a number of bits in the host data. In such systems, the image under investigation is first tested to verify whether it hosts a watermark or not. If the image is indeed watermarked, the embedded message is decoded.

With respect to the method used for watermark detection, techniques can be classified into the following categories:

- *Blind*: The only information required to extract the watermark from the image under investigation is the watermark secret key.
- *Non-blind*: The watermark can be extracted provided that one knows not only the image under test, but also the original image from which it is thought to be derived. Obviously, the requirement that the original image is available during watermark extraction implies serious limitations for the applicability of such algorithms in a number of scenarios. For example,

non-blind techniques cannot be used for the automatic search over the Internet in order to detect illegal copies of images.

In a similar way, watermarking techniques can be distinguished by two major categories with respect to the way embedding is performed:

- *Blind embedding techniques*: Such techniques consider the host image as noise or interference. In most cases, these methods utilize knowledge of the host signal statistics.
- *Informed coding/embedding techniques*: These techniques exploit the fact that, during embedding, not only the statistics of the host image but also the image itself are known, and try to utilize this fact in order to improve watermark detection performance (e.g. through interference cancellation).

Image watermarking has a wide range of applications. A non-exhaustive list of such applications is provided below:

- *Integrity checking, content authentication*: The aim is to detect if an image has been tampered with or not. This is usually achieved by inserting a *fragile* watermark that is destroyed (i.e. it becomes undetectable) if the image is modified. In some cases, content authentication watermarking techniques can provide information about the exact locations of the image that have been altered.
- *Traitor tracing, transaction tracking*: The aim is to trace back a malicious user, e.g. a user that legally possesses an image but has distributed it to other users in an illegal way. This can be done by embedding a robust watermark identifying the customer in each legal image transaction. The term fingerprinting is often used to describe this application.
- *Owner identification, copyright protection*: The aim is to verify that a given document originates from a certain source even after it has been manipulated and tampered with either in an intentional attempt to remove the watermark or during usual operations (e.g. image compression).
- *Usage control*: In this case the embedded information controls the terms of use of the digital content. As an example, the embedded information can be used along with compliant devices to prohibit unauthorized recording of a multimedia document (copy control), or playback of illegal copies (playback control).

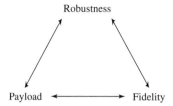

Figure 4.3   Trade off between capacity/payload, visibility/fidelity and robustness

Another fundamental notion in watermarking is the trade off between capacity, visibility and robustness (see Figure 4.3):

- The capacity of a multiple-bit watermarking algorithm is the amount of information, i.e. the length of the message, that can be hidden in the watermarked image. It should be noted that, in most cases, data payload depends on the size of the host data. The more the available host samples, the more bits can be hidden. Thus, capacity is often given in terms of message bits per sample of the host image.
- The term visibility refers to image visual degradation due to the embedding of the watermark. Naturally, a watermark should be as invisible as possible and thus the watermarking process should not introduce suspicious perceptible artifacts. In other words, a human observer should not be able to detect if some digital data have been watermarked or not. The visibility is often expressed quantitatively as the signal-to-noise ratio (SNR) between the marked image and the non-marked image, despite the poor performance of SNR in capturing the way humans perceive distortions.
- Robustness refers to the ability to recover the watermark even after the image has been manipulated and altered in a non-destructive manner (i.e. in a manner that preserves its semantic content, this notion being partly subjective). The alterations can be malicious or not: they can result from common image processing or transmission (i.e. filtering, lossy compression, noise addition) or from an attack attempting to remove the watermark (e.g. the Stirmark attack). Robustness is evaluated by measuring the survival rate of watermarks after attacks.

For a given algorithm, improvements in performance with respect to one aspect usually result in performance degradation with respect to one of the two others.

Numerous still image watermarking algorithms exist in the literature. These algorithms differ in various aspects:

- The selection of the locations in the host image where the watermark is to be embedded: This should be done carefully in order to minimize the distortion or to ensure that the existence of the watermark will remain secret. A direct consequence of Kerckhoff's principle is that the watermarking algorithm should be public and that the watermark should not be 'detectable' in a straightforward manner in order to prevent its removal.
- The choice of the embedding domain: The embedding operation can be performed in the spatial domain, or in a suitable transform domain such as the DCT, Mellin Fourier or wavelet.
- The formatting of the message: Some techniques allow any bit stream to be directly embedded as a watermark, others require a transformation of the message bits prior to the embedding process.
- Embedding procedure: Selection of the rule that will be used for modifying the selected features of the host image so as to embed the watermark. Additive and multiplicative embedding rules are often used.
- Optimization of the watermark detector: Designing the detector module in a way that optimizes watermark detection performance, especially in the case of attacks.

For a more detailed description of image watermarking principles and methods, we refer the reader to Cox *et al.* (2002), Katzenbeisser and Petitcolas (2000) and Tefas *et al.* (2005).

4.3 Watermarking Principles Applied to 3D Objects

4.3.1 Aims of Watermarking

Today, 3D watermarking is a hot topic in the watermarking community. Similar to image watermarking, 3D watermarking aims at hiding in an invisible way information inside a 3D object. This information can then be recovered at any time, even if the 3D object was altered by one or more non-destructive attacks, either malicious or not. Thus 3D watermarking can be useful in several application, security-related ones being the most prominent. For example, users might like to check if the use of a given object is legal or not, to access additional information concerning the object (e.g. for authentication or indexing), the owner (copyright), or even the buyer (e.g. for traitor tracing).

The fact that the creation of 3D models is, in many cases, a labor-intensive and costly procedure made the protection of such models an urgent necessity and attracted the interest of many engineers and researchers to 3D watermarking. As a result, numerous watermarking algorithms for 3D objects have been proposed in the literature.

4.3.2   Trade off Between Capacity, Visibility and Robustness

By analogy with still image watermarking, watermarking of 3D objects involves a trade off between capacity, visibility and robustness:

1. **Capacity**: The amount of information that can be hidden in the 3D object. It should be noted that this amount is closely related to the complexity of the object (e.g. number of vertices, curvature variations) and to the application addressed by the watermarking algorithm (e.g. authentication vs. copyright).
2. **Visibility**: The visual degradation of a 3D object due to watermarking. The visual impact of the watermark on the protected 3D object should be as limited as possible. In order to measure the imperceptibility of the embedded watermark in 3D objects represented as meshes one can use the following metrics:

   (a) *Existing metrics*

   - Hausdorff distance:

   $$H_{max}(M_1, M_2)$$
   $$= \max\{\max_{a \in M_1} \min_{b \in M_2} d(a, b), \max_{a \in M_2} \min_{b \in M_1} d(a, b)\} \quad (4.1)$$

   where M_1 and M_2 are the two meshes and $d(a, b)$ is the Euclidean distance between a and b in the 3D space. This metric is usually called *maximum geometric error*. Another definition of the Hausdorff distance called *mean geometric error* is defined as:

   $$H_{mean}(M_1, M_2) = \frac{1}{A_{M_1} + A_{M_2}} \left(\int_{a \in M_1} \min_{b \in M_2} d(a, b) \right.$$
   $$\left. + \int_{b \in M_2} \min_{a \in M_1} d(a, b) \right) \quad (4.2)$$

The Hausdorff distance is untime to linear transformation and computationally expensive.

- Vertex signal-to-noise ratio (VSNR):

$$\text{VSNR} = \frac{1}{N} \sum_{i=0}^{N-1} d(a_i, b_i) \qquad (4.3)$$

- Geometric Laplacian (GL):

$$\text{GL}(p_i) = p_i - \frac{\sum_{j \in N(p_i)} d(p_i, p_j)^{-1} p_j}{\sum_{j \in N(p_i)} d(p_i, p_j)^{-1}} \qquad (4.4)$$

where $N(p_i)$ is the neighborhood of the vertex p_i and $d(p_i, p_j)$ the Euclidean distance between p_i and p_j. GL differentiates between random noise addition and a compressed version and captures the local smoothness of the mesh, but it usually does not capture the perception of distortions.

(b) *Image-based metric*

This consists of comparing 2D projections of the 3D objects. Lindstrom and Turk (2000) computed root mean square differences of pairs of corresponding 2D views. The main problem with this metric is that perceived degradation of still images is usually different from perceived degradation on the 3D object (experiments of Rogowitz and Rushmeier 2001).

(c) *Evaluation of 3D watermarking perceptual quality*

Gelasca *et al.* (2005) published the first paper or this particular problem. They proposed a roughness-based metric validated through psycho visual experiments.

3. **Robustness**: The ability to recover the watermark even if the watermarked 3D object has been manipulated. The 3D watermarking algorithms exhibit different degrees of robustness to various manipulations of the host 3D objects. Most algorithms are robust to global translation, rotation or scaling of the object, and noise addition to the 3D coordinates of its vertices, whereas fewer algorithms are robust to more challenging attacks like cropping or remeshing (e.g. mesh simplification aimed at reducing the number of triangles to speed up manipulation and rendering).

4.3.3   Embedding Space and Watermark Detection Requirements

One of the main characteristics of a watermarking algorithm is the space or domain in which the watermark is embedded (i.e. workspace). This domain is closely related to the representation used for the object (e.g. mesh, NURBS, etc.), but it may also be a transformed version of this representation. For instance, watermarking of objects represented by 3D meshes can be performed either by modifying the positions of mesh vertices or on the spectral decomposition of the mesh.

Similar to still image watermarking algorithms, 3D object watermarking algorithms can be distinguished by two different classes according to the extraction procedure:

- *Blind algorithms*: Only the private key is needed for watermark extraction.
- *Non-blind algorithms*: To extract the watermark, one should possess not only the 3D object that is to be checked, but also the original 3D object from which it is believed to be derived.

4.3.4   Manipulations and Attacks on 3D Objects

Attacks affecting the data organization within the file

Mesh representations usually consist of a list of vertices and a list of triplets of these vertices, each defining a different triangle. The order of the vertices or triplets within these lists does not affect, in general, the shape of the represented 3D object. This order can thus be manipulated without modifying the 3D object itself (see Figure 4.4). As will be shown later on, some watermarking algorithms use data reordering in order to perform watermark embedding. However, most watermarking algorithms do not modify this order but they rely on the given order for the definition of the embedding positions during watermark insertion and extraction. Obviously, if the vertices or triangles are reordered after watermarking, the watermark extraction becomes impossible and thus this operation can be a relevant attack for this class of algorithms.

Similarly, such degrees of freedom in the organization of the data representing an object may exist with other representations such as NURBS or CSG trees, for instance. Whenever a watermarking algorithm relies on a specific way of data organization within the file, a successful attack consists

OFF			OFF		
49132 98260			49132 98260		
-23.785717	-66.703217	106.301270	-23.785717	-66.703217	106.301270
-27.686848	-64.702660	106.700150	-27.686848	-64.702660	106.700150
-28.791845	-67.008881	106.552475	-28.791845	-67.008881	106.552475
-31.544001	-64.233200	107.175964	-31.544001	-64.233200	107.175964
-32.696758	-67.188759	107.048874	-32.696758	-67.188759	107.048874
-35.629700	-66.258934	107.459938	-35.629700	-66.258934	107.459938
-36.654587	-69.046806	107.307190	-36.654587	-69.046806	107.307190
-39.809319	-65.871834	107.700386	-39.809319	-65.871834	107.700386
-39.393673	-69.070984	107.488113	-39.393673	-69.070984	107.488113
-42.851711	-68.611862	107.594948	-42.851711	-68.611862	107.594948
-40.918461	-71.534988	107.431808	-40.918461	-71.534988	107.431808
-44.594685	-70.812912	107.611198	-44.594685	-70.812912	107.611198
-43.731117	-73.694656	107.549622	-43.731117	-73.694656	107.549622
3 0 1 2			**3 0 2 1**		
3 2 1 3			3 2 1 3		
3 2 3 4			**3 4 2 3**		
3 4 3 5			3 4 3 5		
3 4 5 6			3 4 5 6		
3 6 5 7			3 6 5 7		
3 6 7 8			3 6 7 8		
3 8 7 9			3 8 7 9		
3 8 9 10			3 8 9 10		

Figure 4.4 Example of a manipulation that affects the data organization within a file representing a 3D object. Each triplet of real values defines a vertex, whereas each triplet of integers corresponds to the vertex indices that define a triangular face. The integer 3 indicates the number of vertices per face

of modifying this organization without modifying the 3D object represented by the data.

Attacks affecting the geometric representation

A specific 3D object can be defined in a number of different ways within a given representation principle. For instance, the surface of a 3D object can be meshed in different ways (see Figure 4.5). It is thus possible to alter the mesh representation of a 3D object into another one while preserving its shape (up to some precision level). This attack would be efficient against watermarking algorithms that rely on a given mesh representation to embed and extract the watermark.

The same remark applies to other representations: different NURBS parameterizations can be used to describe the same surface. Thus an algorithm that relies on a given NURBS representation is vulnerable to attacks that change the NURBS parameterization of the object.

OFF

49132 98260

-23.785717	-66.703217	106.301270
-27.686848	-64.702660	106.700150
-28.791845	-67.008881	106.552475
-31.544001	-64.233200	107.175964
-32.696758	-67.188759	107.048874
-35.629700	-66.258934	107.459938
-36.654587	-69.046806	107.307190
-39.809319	-65.871834	107.700386
-39.393673	-69.070984	107.488113
-42.851711	-68.611862	107.594948
-40.918461	-71.534988	107.431808
-44.594685	-70.812912	107.611198
-43.731117	-73.694656	107.549622

3 0 1 2
3 2 1 3
3 2 3 4
3 4 3 5
3 4 5 6
3 6 5 7
3 6 7 8
3 8 7 9
3 8 9 10

OFF

49132 98260

-23.785717	-66.703217	106.301270
-27.686848	-64.702660	106.700150
-28.791845	-67.008881	106.552475
-31.544001	-64.233200	107.175964
-32.696758	-67.188759	107.048874
-35.629700	-66.258934	107.459938
-36.654587	-69.046806	107.307190
-39.809319	-65.871834	107.700386
-39.393673	-69.070984	107.488113
-42.851711	-68.611862	107.594948
-40.918461	-71.534988	107.431808
-44.594685	-70.812912	107.611198
-43.731117	-73.694656	107.549622

3 0 5 2
3 0 1 3
3 1 3 4
3 1 2 4
3 2 3 5
3 6 5 7
3 6 7 8
3 8 7 9
3 8 9 10

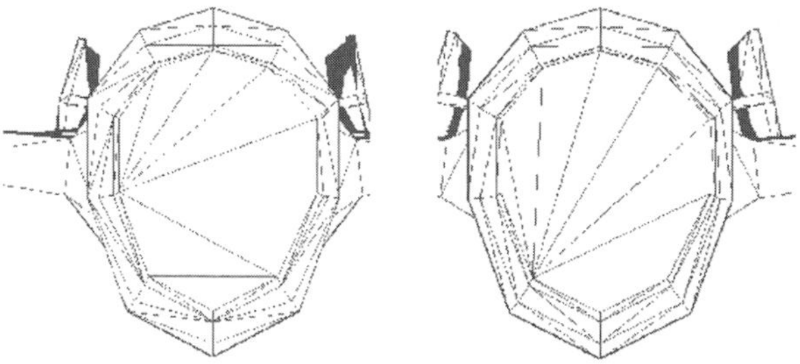

Figure 4.5 Manipulations affecting the representation of the 3D object

Noise on geometry

Assuming that the watermark is encoded through minor modifications of the geometrical entities used to define a 3D object, e.g. by slightly altering the location of the vertices of a mesh, then the watermark may be affected by noise applied on the geometry (see Figure 4.6). Such noise could be introduced either maliciously to weaken the watermark, or in the course of typical 3D object manipulations like lossy compression, or format conversion.

Figure 4.6   Random noise applied to the geometry of a 3D mesh with different magnitude

Global transformations

A watermarked object can be subject to an affine transform like translation, rotation, uniform or non-uniform scaling, or even a projective transform or another non-affine global transform (see Figure 4.7). Some algorithms rely on the precise position and orientation of an object to extract the watermark. In this case, a global transformation applied, maliciously or not, to the object can hamper watermark detection unless a way is found to recover the object's original reference frame used for watermarking.

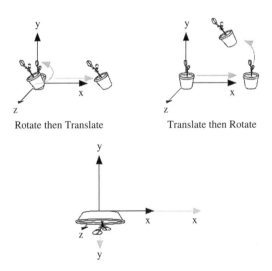

Figure 4.7   Global transformations applied on a 3D object

Figure 4.8   Cropping of geometry

Cropping of geometry

Similar to images, it is possible to remove part of an object's geometry (see Figure 4.8). In some cases this would yield a meaningless object, but in other cases the remaining part could still be of some value, even though the watermark may be destroyed in the process. In fact, cropping will often destroy the structure of the watermark embedding space (e.g. the ordering of vertices), unless the watermarking algorithm is designed to be robust to cropping, which may involve hiding the watermark several times in different parts of the object.

Mesh simplification attacks

The goal of mesh simplification is to speed up the manipulation and rendering of 3D objects. It consists of displaying the 3D polygonal mesh with fewer triangles while preserving the same shape (see Figure 4.9). It is a challenging attack for the 3D watermarking community.

Readers are invited to refer to the compression chapter (Chapter 2) for more details about this attack.

Mesh smoothing

Meshes obtained from real-world objects are often noisy. Mesh smoothing relocates mesh vertices to improve the mesh quality while keeping the mesh topology unchanged (see Figure 4.10). Several techniques can be used for mesh smoothing; we can cite the umbrella-operator smoothing (see Figure 4.11) and Taubin's lambda–mu smoothing.

Figure 4.9   Mesh simplification

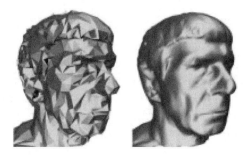

Figure 4.10   Mesh smoothing in a 3D mesh: left, before smoothing, right, after smoothing

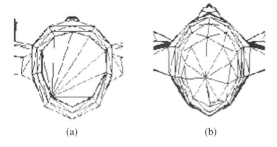

(a)   (b)

Figure 4.11   Impact of the umbrella-operator smoothing in a 3D mesh: (a) before smoothing and (b) after smoothing

Generic attacks

Watermarking of 3D objects is subject to generic attacks on watermarking protocols. Such attacks include collusion and de-watermarking.

De-watermarking attacks are methods that try to remove the watermark by taking advantage of the fact that the watermarking algorithm (especially its embedding space and its weaknesses) is known.

On the other hand, a collusion attack involves several people that possess the same watermarked object, hosting different watermarks, and cooperating in order to remove or detect the watermark. In fact, by simply averaging the various watermarked instances of the object into a single object, the different watermarks might be averaged (due to their random nature) into a weak noise, while the object itself would remain intact. Another collusion attack consists of combining pieces from various watermarked instances of the same object so as to obtain a complete object featuring pieces from different watermarks. It would thus be impossible to detect any complete watermark in the resulting composite object.

4.4   A Guided Tour of 3D Watermarking Algorithms

In this section, we briefly describe the most cited algorithms for 3D models in order to provide a review of the state of the art in 3D watermarking algorithms.

The reviewed algorithms are classified according to two criteria.

The first criterion is the method used to embed the watermark in the object. According to this criterion, watermarking algorithms are classified in the following categories:

1. **Data file organization**: These algorithms encode information by modifying the organization of the data in the computer file associated with the 3D object.
2. **Topological data**: These algorithms, which operate on mesh data, use the topology of the 3D object, i.e. the connectivity of the mesh, to embed data. The geometry of the mesh, i.e. positions of vertices, is not modified.
3. **Geometrical data**: These algorithms are based on slight modifications performed on the geometric data of the 3D object. This category is the largest one and has been organized by starting with algorithms that operate in the spatial domain followed by approaches using some multiresolution representation and then techniques that work in the spectral domain.

The second criterion used for our classification of 3D object watermarking algorithms is the representation used to define the 3D object. The following main classes are used in the classification presented overleaf:

1. Mesh.
2. NURBS.
3. Others.

4.4.1   Data File Organization

This category includes only three algorithms, each operating on a different 3D object representation scheme, namely 3D meshes, NURBS and CSG trees.

Algorithm 1 *Ichikawa* et al. *(2002): mesh, reordering of vertices or triangles*

This algorithm encodes information in a mesh represented by a list of vertices and a list of triangles. To achieve this, it modifies, in the computer file, the order of the triangles in the list of triangles, or the order of the triplet of vertices forming a given triangle. This algorithm does not modify either the geometry or the topology of the mesh.

Algorithm 2 *Ohbuchi and Masuda (n.d.), Ohbuchi* et al. *(1999): NURBS, reparameterization*

These algorithms hide data in a NURBS curve or surface by changing the parameter involved in the NURBS parameterization through a homographic function. Such a change of parameter has three degrees of freedom that are reduced to one by imposing some constraints. The remaining degree of freedom is used to encode the hidden information. Changing the parameter also involves changing the nodes and weights accordingly in order to preserve the geometry. Thus, these algorithms do not modify the geometry.

Algorithm 3 *Fornaro and Sanna (2000): CSG tree*

This algorithm aims at hiding information in a 3D model described by a CSG tree. For this, new kinds of nodes for the CSG are defined: the watermark (control) nodes. Nodes in this new category are linked to the original CSG and contain the watermark information. To achieve invisible watermarking, control nodes are null-volume objects (e.g. null-radius spheres). The information that is hidden in those nodes is a hash value of the original 3D model, encrypted by the secret key of an asymmetric cryptographic algorithm. The targeted application is thus the authentication of the object that is achieved by comparing the computed hash value of the object under investigation to the decrypted hash value hidden in the new nodes.

4.4.2   Topological Data

This class of algorithms uses the topology (connectivity) of the 3D object to insert the watermark. All algorithms reviewed in this section operate in 3D objects defined as a 3D mesh.

Algorithm 4 *Ohbuchi* et al. *(1997): mesh, 'Triangle Strip Peeling Symbol (TSPS) sequence embedding'*

This algorithm uses the topology of the polygonal 3D model but does not modify its geometry. Starting from an initial edge of the mesh, a strip of triangles is defined by successively attaching a new triangle to the current strip as follows: the last triangle of the current strip has two 'free' edges (i.e. edges that are not connected to the strip) each corresponding to a triangle that is not yet attached to the strip. The selection of the next triangle to be attached to the strip among the two possible choices is based on the next bit of the message to be hidden. Once the strip is defined, it is 'peeled off' by duplicating all edges and vertices on the boundary of the strip except from the starting edge. Thus the strip is connected to the rest of the mesh only by the starting edge (see Figure 4.12). This 'peeling' of the strip changes the topology but is invisible since its boundary edges and vertices coincide with those defining the corresponding boundary on the remainder of the mesh. For message extraction, the starting edge of the strip is found and, as the strip is traversed, the embedded bits are extracted based on the strip's path.

Algorithm 5 *Ohbuchi* et al. *(1997): mesh, 'Polygon Stencil Pattern (PSP)'*

Given a 3D object described by a mesh, a visual pattern (e.g. a letter) can be hidden in it by 'peeling off' a strip of triangles that has the shape of the pattern to be hidden (e.g. a strip that forms the letter to be hidden (see

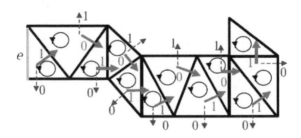

Figure 4.12   A strip of triangles is selected to insert the message '10101101011'

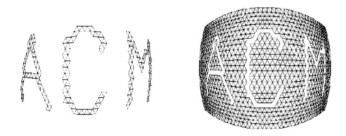

Figure 4.13 A 3D object watermarked using the 'Polygon Stencil Pattern' technique

Figure 4.13)). Just like Algorithm 4, the strip is 'peeled off' by duplicating the edges and vertices that form the boundary of the strip. Thus, only the topology of the mesh is modified, not the geometry. Furthermore, when the 'peeled' strip of triangles is displayed along with the rest of the mesh, no alterations are visible. This algorithm is somewhat robust to remeshing.

Algorithm 6 *Ohbuchi* et al. *(1997): mesh, 'Mesh Density Pattern embedding'*

This algorithm locally modifies the density of triangles in a mesh so as to construct a visual pattern, observable when viewing the 3D object in wireframe mode (see Figure 4.14). This visible watermark is robust to geometrical operations of rotation, scaling and translation (RST) and is resistant but not immune to polygonal simplification and other topological manipulations.

Figure 4.14 A 3D object watermarked using the 'Mesh Density Pattern embedding' technique

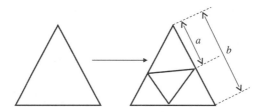

Figure 4.15   The watermark is embedded in the ratio a/b

Algorithm 7 *Mao* et al. *(2001): mesh, triangle subdivision*

This algorithm retriangulates a part of a triangular mesh and embeds the watermark into the positions of the newly added vertices. Triangles are chosen according to a secret key in such a way that the ratio of two line segments lying on the same straight line encodes the hidden information (see Figure 4.15). The embedded watermark can be extracted only from the stego-model without using the original cover model. This algorithm achieves high capacity and is robust to affine transformations, whereas it is easily destroyed by local deformations and topological alterations.

4.4.3   Geometrical Data

The majority of the 3D watermarking algorithms insert the watermark by modifying the geometry of the 3D object. Most techniques deal with mesh data. In this category, some of them operate in the spatial domain by modifying vertices, normals (direction, length) and geometrical invariants (i.e. length of a line, area of a polygon, etc.). Others embed information in a transform domain: spectral decomposition, wavelet transform and spherical wavelet transform.

Watermarking algorithms operating on 3D models represented by NURBS, point sets or other forms of representation are less widespread.

In this section, we present a brief description of the best known techniques.

Mesh representation

Spatial domain

Algorithm 8 *Bors (2004a, 2004b, 2006), Harte and Bors (2002): mesh, vertex displacement*

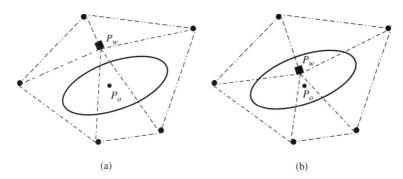

(a)   (b)

Figure 4.16 P_o and P_w are respectively the original and watermarked vertices. The configuration (a) embeds a bit '0' and the configuration (b) embeds a bit '1'

This watermarking algorithm is based on the position of the vertices and not on their connectivity. It consists of two steps. In the first step, a list of vertices and their neighborhoods are selected from mesh areas consisting of small polygons and are ordered according to their Euclidean distance from the vertex that is the head of the list (i.e. the vertex with the smallest squared distance to its neighbors). In the second step, locations of selected vertices are changed according to their local moments and the information bit to be embedded (see Figure 4.16). The watermark can be recovered after scaling, rotation or a combination of geometrical transformations. Robustness to various levels of noise added on vertices as well as to 3D object cropping was experimentally verified.

Algorithm 9 *Ashourian and Enteshary (2003): mesh, vertex displacement*

This non-blind watermarking algorithm operates in the spatial domain. It inserts the watermark signal (W_{xi}, W_{yi}, W_{zi}) into the vertex coordinates (X_i, Y_i, Z_i) using the following equations:

$$\begin{cases} X_i' = X_i + a \cdot M_x(p_i) \cdot W_{xi} \\ Y_i' = Y_i + a \cdot M_y(p_i) \cdot W_{yi} \\ Z_i' = Z_i + a \cdot M_z(p_i) \cdot W_{zi} \end{cases}$$

with the masking functions $(M_x(p_i), M_y(p_i), M_z(p_i))$ defined as the difference in coordinate values of each point with the points connected to it.

This algorithm is robust to additive noise, geometrical compression performed by the MPEG-4 SNHC standard and mesh simplification.

Algorithm 10 *Barni* et al. *(2004): mesh, vertex displacement*

This algorithm perturbs vertices positions of the 3D model according to a spherical pseudo-random bumped surface. The pseudo-random position and amplitude of the bumps encode the watermark. Watermark recovery is achieved via a standard correlation detector. This algorithm can achieve a good degree of robustness in the case of 3D objects with a fairly large number of faces.

Algorithm 11 *Yeung and Yeo (1998), Yeo and Yeung (1999): mesh, fragile watermark*

To check the integrity of a triangular mesh, Yeung and Yeo have developed a blind fragile watermarking algorithm that slightly moves each vertex.

The insertion process is as follows:

1. Compute a double position index $L = (L_x, L_y)$ for each vertex v.

   - Define the centroid s of the set of vertices adjacent to v:

$$s = \frac{1}{|N(v)|} \sum_{u \in N(v)} u \qquad (4.5)$$

   .

   - Convert s coordinates into integers (N_x, N_y, N_z).
   - Combine these integers to obtain L_x and L_y. An example combination is $L_x = (N_x + N_y + N_z)$ mod XSIZE.

2. Compute a triple value index $p(v) = (p_1, p_2, p_3)$ for each vertex v:

   - Convert v's coordinates into a triplet of integers.

3. Generate a binary sequence using a secret key K and values of indices $p(v)$. Each $p(v)$, given as input to a conversion table parameterized by K, yields a binary value $K(p(v))$.

4. Slightly modify the geometry of the model (i.e. the vertices v) so that the bit $W(L(v))$ of the watermark W (defined as a black and white 2D image) is equal to $K(p(v))$.

The verification step consists of evaluating $K(p(v))$ and $W(L(v))$ for each vertex v, and then computing a correlation score c between the two binary sequences:

$$c = \frac{|\{v : K(p(v)) \neq W(L(v))\}|}{|\{v\}|} \qquad (4.6)$$

If the object has not been modified, the correlation score c is equal to 1. This method is not robust to vertex renumbering due to the definition of the centroid s in the insertion step.

Algorithm 12 *Benedens (1999b), Benedens and Busch (2000): mesh, 'Vertex Flood'*

This high-capacity method hides a watermark in a 3D object represented by a set of triangles. The algorithm, which affects only the geometry of the object, modifies the distance between vertices of the model and the center of gravity of a given reference triangle for each selected interval. In the retrieval process, the mean distance of all vertices included in an interval can be used for decoding the embedded bits. This method can increase robustness to randomization of simple vertices.

Algorithm 13 *Yu et al. (2003), Zhi-qiang et al. (2003): mesh, vertex displacement*

This algorithm is an extension of Algorithm 12. It partitions the vertices of a mesh into N sets, pseudo-randomly defined from a secret key. The watermark is encoded in the distance between a vertex and the barycenter of the set it belongs to, as follows:

$$L_{oij}^{w} = L_{oij} + \alpha W_i U_{oij} \tag{4.7}$$

where L_{oij} denotes the original vector from the center of the model to the jth vertex of the ith section and U_{oij} denotes the vector whose direction is the same as that of L_{oij} and its amplitude is the minimal length of the jth vertex's 1-ring edge neighborhood. The watermark retrieval is achieved by first computing an estimation of the watermark. For this, the algorithm evaluates for each set the difference in length between the vectors of the original model that link the vertices to the center of the set and the vectors of the detected model that also link the vertices to the set center. Then the correlation coefficient between the extracted watermark sequence and the designated watermark sequence can provide information about the watermark presence or absence. Experiments show that this method is robust to many attacks, especially cropping and noise, and is made robust to remeshing by the inclusion of a registration/resampling step that aims at aligning the mesh under investigation with the original mesh before watermark extraction.

Algorithm 14 *Benedens (1999b), Benedens and Busch (2000): mesh, 'Triangle Flood'*

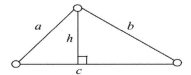

Figure 4.17   Dimensionless quantities for a triangle: a/b and h/c

This is another high-capacity algorithm. The targeted applications are the same as for Algorithm 12. The algorithm first generates an order for walking over the set of triangles, and then hides information in the heights of the successive triangles by slightly modifying the positions of their vertices.

Algorithm 15 *Ohbuchi* et al. *(1997): mesh, 'Triangle Similarity Quadruple (TSQ)'*

This algorithm is based on the modification of the geometry of a triangular mesh. For each triangle a couple of dimensionless quantities are considered (see Figure 4.17). These quantities may be for example the length ratio of two edges of a triangle (a/b) or the length ratio between one edge and the corresponding height (h/c). An integer value is encoded into a triangle by slightly modifying the associated quantities.

In fact the building block, Macro Embedding Primitive (MEP), for encoding information is a set of four adjacent triangles (quadruple) as shown in Figure 4.18. The integer value hidden in one of the four triangles (marker) identifies the MEP. Two other triangles encode a part of the payload (data 1 and data 2). The last triangle encodes a sequence number, indicating how to assemble the payload pieces obtained from the various quadruples into the original complete message (subscript). The extraction of the watermark consists of traversing the triangular mesh and finding all the triangles with the marker. Then the two data symbols and the subscript are extracted from the triangles in the MEP.

Algorithm 16 *Ohbuchi* et al. *(1997): mesh, 'Tetrahedral Volume Ratio (TVR) embedding'*

This algorithm hides a piece of information in a triangular mesh by modifying an affine-invariant value of a pair of tetrahedra. This affine invariant is the ratio of their volumes. A tetrahedron is simply defined by an edge and its two adjacent triangles.

Algorithm 17 *Cayre and Macq (2003): mesh, variant of TSPS*

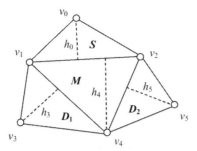

Figure 4.18   Macro Embedding Primitive for the TSQ watermarking technique. The marker (M) is embedded by slightly changing the coordinate of vertices v_1, v_2 and v_4 such that $\{e_{14}/e_{24}, h_4/e_{12}\} = \{b/a, h/c\}$ with e_{ij} the distance between vertices v_i and v_j. The subscript (S) is coded in the pair $\{e_{02}/e_{01}, h_0/e_{12}\}$ by slightly changing the coordinates of vertex v_0. Data 1 (D_1) and data 2 (D_2) are coded in pairs $\{e_{13}/e_{34}, h_3/e_{14}\}$ and $\{e_{45}/e_{25}, h_5/e_{24}\}$ by slightly changing the coordinates of vertices v_3 and v_5

This algorithm considers that each triangle is a two-state variable. The state of a triangle is determined by the position of the projection of one of its vertices on the opposite edge. This state can be modified by slightly moving the relevant vertex as shown in Figure 4.19. The set of triangles that encode the binary message is defined as a strip of triangles by a procedure analogous to that of Algorithm 4 (TSPS): an initial triangle is chosen and one moves from this triangle to one of the two possible adjacent triangles, as dictated by a pseudo-random binary sequence. The difference with the TSPS algorithm is that in this algorithm the hidden data are not encoded in the shape of a strip of triangles. Furthermore, the strip of triangles is not cut out of the object's surface, thus the topology is not modified.

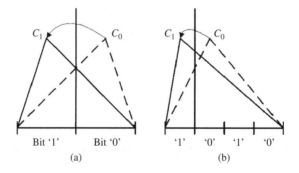

Figure 4.19   A bit '1' is encoded in both cases (a) and (b)

Note: Watermarking algorithms that operate in the spatial domain via slight modification of the vertices' coordinates generally are not robust to noise addition.

Algorithm 18 *Wagner (2000): mesh, length of normals*

This algorithm embeds the watermark in the length of the 'normals' n_i defined on each vertex. The normal at a vertex is defined as the norm of the vector between the barycenter of the vertex's neighbors and the vertex itself. Modifying these normals by replacing some bits of n_i with the bits of the watermark implies moving the vertices. The watermark extraction can be achieved by calculating the value of the normal vector length and extracting the appropriate bits in which the watermark is hidden. The use of a norm invariant to affine transformations yields a watermark that is robust to affine transforms of the 3D object. The paper by Maret and Ebrahimi (2004) proposes an extension that increases the embedding capacity.

Algorithm 19 *Benedens (1999a, 1999c), Benedens and Busch (2000): mesh, direction of normals*

The watermark embedding space used in this algorithm is the direction of the normals of the object triangles. The unit sphere, representing the set of all oriented directions, is divided into regions. The normal of each triangle belongs to one of these regions. The dispersion of the set of normals belonging to a given region of the unit sphere is actually used to encode a bit. To encode a zero the dispersion of normals in a given region is decreased; to encode a one it is increased. This is achieved by altering the normals and ultimately the vertices of the object. The regions of the unit sphere that are chosen to encode a bit are determined by a secret key.

Algorithm 20 *Kwon* et al. *(2003): mesh, direction of normals*

This algorithm is an improvement of Algorithm 19. The embedding space is the histogram of the normals' direction on the unit sphere. The algorithm achieves robustness against cropping by hiding the watermark in several parts of the object. Furthermore, the authors claim that the algorithm is robust to remeshing and isometries.

Algorithm 21 *Song and Cho (2004), Song* et al. *(2002): mesh, cylindrical depth map*

This algorithm computes a cylindrical depth map associated to a mesh (see Figure 4.20). The cylindrical frame of reference is aligned on the principal

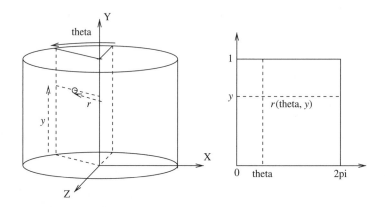

Figure 4.20   Computing the depth map of a 3D object

axis of inertia of the 3D object. The depth map is watermarked using a 2D image watermarking algorithm and the modifications of the depth map are applied back to the original mesh object.

Algorithm 22 *Kalivas* et al. *(2003): mesh, principal component analysis*

This algorithm is based on the still image watermarking technique originally proposed by Tefas and Pitas (2001b). Robustness to translation is achieved by translating the model so that its center of mass falls on the origin of the coordinate system. To achieve robustness against rotation, the model is rotated so that the principal component of the vertices (the eigenvector that corresponds to the greatest eigenvalue of the covariance matrix of vertex coordinates) coincides with the z axis. These operations are performed before both watermark embedding and detection. Subsequently, the model vertices are represented in spherical coordinates (r, θ, ϕ), ordered according to whether their θ value and their r value (distance from the center of mass) are modified according to the watermark bits. Essentially this procedure corresponds to performing the embedding on the 1D signal $r(\theta)$. The fact that only the r value is changed by the embedding process makes the method robust to uniform scaling.

Algorithm 23 *Zafeiriou* et al. *(2005): mesh, SPOA (Sectional Principal Object Axis) watermarking*

This algorithm is an improvement of Algorithm 22. Instead of using one vertex for the embedding of a bit of the watermark sequence, this method embeds a watermark bit in multiple vertices. More specifically, the model is translated and rotated so that its center of mass coincides with the origin of the coordinate system and the principal component is aligned with the z axis. Then vertices are transformed to spherical coordinates (r, θ, ϕ) and the range of θ values is split into a number of consecutive non-overlapping intervals. The vertices whose θ values lie in two consecutive intervals are used to embed one watermark bit. This is achieved by modifying the r values so as to enforce certain constraints. Detection is performed by checking the validity of these constraints on the model under investigation. The proposed algorithm is robust to rotation, translation, uniform scaling, noise addition and mesh simplification, but is vulnerable to cropping.

Algorithm 24 *Koh and Chen (1999): mesh, progressive transmission*

This algorithm aims at watermarking the stream of data used to send a mesh in a progressive manner over a network. Each vertex is transmitted one by one, in a 'progressive' order. The watermark is in fact embedded in the 1D signal consisting of the sequence of transmitted vertices. It is possible to add a pseudo-random sequence to the 1D signal to get it watermarked. The watermark is detected upon reception of the data stream. The correlation coefficient between the embedded watermark and the transmitted signal indicates if the stream is watermarked or not (i.e. it is a zero-bit watermarking algorithm).

Transform domain

Algorithm 25 *Kanai* et al. *(1998): mesh, wavelet decomposition*

This algorithm performs a wavelet decomposition of the mesh. The watermark is then embedded in the wavelet coefficients, and the watermarked mesh is obtained by reconstruction from the modified wavelet coefficients. Only the wavelet coefficients of significant magnitude are modified. The algorithm is based on the notion of spread spectrum (see Cox *et al.* 1997).

Algorithm 26 *Uccheddu* et al. *(2004): mesh, wavelet decomposition*

This blind watermarking algorithm is based on wavelet decomposition of semi-regular meshes. The watermark is embedded in the wavelet coefficients of a suitable resolution level. Watermark detection is accomplished by computing the correlation coefficient between the watermark and the mesh under inspection. Robustness against geometrical transformations (rotation, translation, uniform scaling) is achieved by embedding the watermark in a normalized version of the mesh, obtained by means of PCA (Principal Component Analysis). This technique has been extended to irregularly subdivided 3D triangular meshes (Valette *et al.* 1999) by using lazy wavelets defined on arbitrary connectivities, but the algorithm still fails after connectivity attacks.

Algorithm 27 *Praun* et al. *(1999): mesh, wavelet decomposition*

This algorithm embeds a watermark in wavelet coefficients computed by the progressive representation of a mesh introduced by Hoppe (1996) and associates a function to each vertex duplication operation. This function defines a deformation of the object by moving the vertices neighboring the duplicated vertex. For inserting the watermark, the duplication operations that result in the biggest modifications of the object's shape are chosen, and the associated deformation functions are used and weighted by the watermark information to deform the object slightly. The extraction of the watermark is done in a non-blind manner after registering and resampling the tested object with respect to the original object.

Algorithm 28 *Yin* et al. *(2001): mesh, wavelet decomposition*

The authors of this algorithm aim at generalizing the idea of Algorithm 27 and applying it on other multiresolution representations of a triangular mesh. The underlying principle is, however, somewhat different: a multiresolution representation of the mesh is constructed and then the watermark is encoded by modifying the shape (via small displacements on a subset of the vertices) of a certain, properly chosen, resolution level (instead of using several resolution levels as in Algorithm 27). The watermark extraction is performed after registering and resampling the watermarked object with respect to the original object, so as to be robust to attacks such as format conversion.

Algorithm 29 *Jin* et al. *(2004): mesh, spherical wavelet transform*

The basic idea of this non-blind algorithm is to decompose the original mesh into some detailed parts and an approximation part by using spherical wavelet transformation. The watermark is embedded into both the detailed and the approximation parts. The embedding process starts with global spherical parameterization, spherical uniform sampling of the host mesh, and spherical wavelet forward transformation. Once in the wavelet domain, watermark embedding is performed, the inverse transform is applied and the watermarked mesh is resampled to recover the connectivity of the original model. The watermark detection process includes alignment of the watermarked mesh with the original one, global spherical sampling, spherical wavelet forward transformation on both the original and the watermarked model and finally watermark extraction through comparison of these two meshes. This algorithm is robust to reordering of vertices, mesh simplification and noise addition.

Note: Watermarking algorithms operating in the wavelet domain offer a good control of the local distortion caused by the embedding process.

Algorithm 30 *Ohbuchi* et al. *(2002): mesh, spectral decomposition*

This algorithm operates in the mesh 'spectral domain', obtained after diagonalizing the combinatorial Laplacian matrix of the mesh. Diagonalizing a large matrix (whose dimension equals the number of vertices) is a very heavy and numerically unstable operation. That is why the mesh is first divided into smaller regions which are processed independently. The watermark message is duplicated many times and is used to modulate the amplitude of the spectral coefficients. To extract a watermark, the algorithm compares the shape of the reference (i.e. original) mesh to the watermarked (and possibly attacked) mesh in the mesh spectral domain. The extraction requires information on how the original mesh is partitioned. This non-blind algorithm is robust to several attacks. Prior to detection, a registration/resampling step with respect to the original model is performed to ensure robustness to remeshing.

Algorithm 31 *Cayre* et al. *(2003): mesh, spectral decomposition*

This algorithm is very similar to Algorithm 30. The Laplacian matrix used to obtain the frequency decomposition is the one originally proposed by Taubin *et al.* (1996) and not the combinatorial Laplacian used in Algorithm 30. The difference with respect to Algorithm 30 is that each region subject to spectral decomposition is first resampled to obtain a given regular connectivity graph for the vertices (typically, the valence of each vertex is 6 except at the edges). Since this connectivity graph is known in advance, the frequency decomposition matrix can be pre-computed. Another difference with Algorithm 30 is that it does not perform a registration/resampling step before extracting the watermark. This makes the algorithm non-robust to remeshing, but this is only a 'strategic' choice, unrelated to the embedding technique used by the algorithm.

Algorithm 32 *Alface and Macq (2006): mesh, spectral decomposition*

This blind algorithm proceeds by first partitioning the mesh using a geodesic Delaunay triangulation of a number of feature points automatically selected through a multi-scale estimation of the curvature tensor field. Then, each of the geodesic triangle patches is parameterized and remeshed so as to generate a robust base mesh. Finally, the remeshed patches are watermarked in the mesh spectral domain following the work of Cayre *et al.* (2003) (Algorithm 31). This algorithm is robust to affine transformations, cropping and connectivity attacks.

Algorithm 33 *Wu and Kobbelt (2005): mesh, spectral decomposition*

This algorithm is very similar to Algorithm 30. The major difference is that it uses a set of geometry-dependent orthogonal basis functions derived from radial basis functions (Carr *et al.* 2001; Ohtake *et al.* 2004) to span the spectral space rather than using Laplace basis functions which emerge from the Laplacian matrix. In that way, this method runs faster and can thus efficiently watermark very large meshes.

Algorithm 34 *Murotani and Sugihara (2003): mesh, singular spectrum analysis (SSA)*

This algorithm adds a watermark into a 3D polygonal mesh in the spectral domain. The 3D polygonal model is considered as a sequence of vertices called a vertex series. The spectra of the vertex series are computed using SSA (Galka 2001; Golyandina *et al.* 2001) for the trajectory matrix derived from the vertex series. The watermark is added in the spectral domain by modifying the singular values. The watermark can be extracted by comparing in the spectral domain the singular values of the watermarked and the original data. This non-blind watermarking algorithm is robust against similarity transforms and moderate noise added to vertex coordinates. An improvement of this algorithm has been proposed by Murotani and Sugihara (2004) to achieve robustness against random noise.

Note: Spectral decomposition provides very good robustness against attacks. Spectral decomposition watermarking algorithms are promising. The main weakness is the computational complexity (e.g. the diagonalization of a large matrix whose dimension equals the number of vertices is a very heavy and numerically unstable operation).

Algorithm 35 *Daras* et al. *(2004), Jin* et al. *(2004): mesh, generalized Radon transform*

Watermarks generated by this algorithm are to be used for content-based indexing and retrieval of 3D models stored in a database. The proposed approach is based on the use of a generalized Radon transform. More precisely, a cylindrical integration transform (CIT) is applied to the 3D model to produce descriptor vectors. Each descriptor vector corresponds to a cylinder and furthermore to a set of vertices lying inside the cylinder. The watermark is embedded via a modification of the location of these vertices according to a unique sequence of bits which is used as an identifier linking each model to its descriptor vector.

Algorithm 36 *Jeonghee* et al. *(2003): mesh, discrete cosine transform (DCT)*

This algorithm operates in the DCT domain. The 3D mesh is traversed to generate strips of triangles and transform their vertex coordinates into frequency coefficients in the DCT domain. The watermark is then embedded into the mid-frequency band of AC^1 coefficients for robustness and imperceptibility. The extraction of the watermark is performed via a comparison of the mid-frequency coefficients of the original and watermarked 3D model. Experiments show that the inserted watermarks survive various attacks, such as additive noise, geometry compression, affine transformation and multiple watermarking.

Other representations

Algorithm 37 *Benedens (2000): NURBS*

This algorithm uses the same principle as Algorithm 16 while introducing some small improvements. Its most interesting feature is that it uses a mesh watermarking algorithm to watermark 3D objects represented by NURBS. For this, the NURBS surface is first tessellated, then the mesh watermarking algorithm is applied to the obtained mesh, and finally the deformations induced on the mesh by the watermark are applied to the original NURBS coefficients.

Algorithm 38 *Lee* et al. *(2002): NURBS, orthographic depth maps*

This algorithm computes three orthographic depth maps of a NURBS object (see Figure 4.21) and then watermarks them using a 2D image watermarking algorithm. The modifications of the depth maps induced by the 2D watermarking process are applied back to the original NURBS control points. The selection of the reference frame used to compute the depth maps is controlled by a secret key.

---

[1] There names come from the historical use of DCT for analyzing electric circuits with direct and alternating currents.

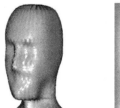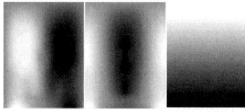

Figure 4.21   A 3D model and the corresponding three orthographic maps

Algorithm 39 *Mitrea* et al. *(2004): NURBS, spread spectrum*

This NURBS surface watermarking algorithm operates on 2D DCT coefficients by means of a spread spectrum procedure. Firstly, three virtual images are computed from the NURBS representation: the pixel values in these images are the coordinates of the NURBS control points. A 2D DCT is then applied to each image to obtain a vector of characteristics. Then, the secret information (key) and public information (logo) are combined by means of a code division multiple access (CDMA) technique to provide the watermark which is subsequently embedded into the vector of characteristics by means of a weighted addition. Watermark detection is achieved by means of matched filters.

Algorithm 40 *Louizis* et al. *(2002): voxels, watermarks of specific spatial structure*

This algorithm is based on the image watermarking technique proposed by Tefas and Pitas (2001b) and, unlike most algorithms, operates on volumetric (voxel-based) models. A watermark signal of specific structure (having a self-similar nature) is embedded in the spatial domain and enables blind, fast and robust progressive watermark detection even after geometric distortions of the watermarked volume. The term progressive watermark detection means that under certain conditions the watermark detection procedure does not have to be performed on the entire volume; the volume is scanned sequentially until a decision on the existence of a watermark can be reached.

Algorithm 41 *Ohbuchi* et al. *(2004): point set*

This non-blind algorithm is applicable to 3D objects defined as a point cloud, i.e. a set of points that are not connected by edges. It follows the same frequency-domain shape modifications approach as the 3D mesh–

watermarking algorithm by Ohbuchi *et al*. (Algorithm 30). For the watermark embedding, the algorithm creates a disjoint set of clusters from the point set, generates a mesh for each cluster of points and applies mesh spectral analysis on each mesh. Then it modifies the mesh spectral coefficients according to the information bit string to be embedded, performs an inverse transform of the coefficients back into the vertex coordinates domain and finally discards the vertex connectivity of the mesh to obtain a watermarked point set. For watermark extraction, the watermarked point set is aligned with the original one, and its geometry is resampled using the original one. Then the clusters on the reference point set are recreated and the clustering is transferred onto the resampled watermarked one. Subsequently a mesh is created for each cluster, mesh spectral analysis is performed on both the original and watermarked point sets and the embedded bit string is extracted by comparing the coefficients of the corresponding clusters.

4.4.4 Others

Algorithm 42 *Ohbuchi* et al. *(1998a, 1998b): miscellaneous, attributes*

These papers refer to techniques that achieve watermarking of 3D objects through the modification of certain attributes attached to the object's geometry, such as texture coordinates or color/opacity parameters associated with each vertex, line or face. In Ohbuchi *et al*. (1998a) preliminary results we presented for a watermarking algorithm that operates in the texture coordinates. Other attributes mentioned for possible watermark embedding are VRML scene parameters or 3D animation tables for deformable models.

Algorithm 43 *Hartung* et al. *(1998): miscellaneous, animation parameters*

This algorithm belongs only marginally to the topic covered by this chapter since it involves synthetic video sequences of 3D face models, animated from a stream of MPEG-4 facial animation parameters. What is watermarked here is not the 3D object but the parameter stream. For the embedding of watermark data a spread spectrum approach was adopted.

Algorithm 44 *Garcia and Dugelay (2003): miscellaneous, asymmetric 3D/2D watermarking procedure*

This algorithm for watermarking 3D textured objects is based on the object's texture map instead of its geometrical data. The main goal of the algorithm is to hide information in the texture image in such a way that one can retrieve hidden information from images or videos generated from the 3D synthetic object, thus protecting the visual representations of the object.

Given a 3D object consisting of its geometry, a texture image and a texture mapping function, information is embedded in the object by watermarking its texture image via a still image watermarking algorithm. Watermark detection is performed on views (i.e. 2D projections) of the object and is achieved by reconstructing the watermarked texture and extracting the watermark from the recovered texture image.

Algorithm 45 *Bennour and Dugelay (2006): miscellaneous, asymmetric 3D/2D watermarking procedure*

Similar to Algorithm 44, the main goal of this framework is to retrieve the watermark from images or videos resulting from the 3D synthetic object, thus protecting the visual representation of the object. However, this framework is based on the object's apparent contour instead of its texture map. Thanks to this point, this approach can be used to protect the images derived (i.e. projected) from a watermarked 3D object with or without texture.

Given a 3D object, its 3D silhouette is extracted, sampled and watermarked using a robust algorithm designed for 3D polygonal lines in order to get the watermarked 3D object. This object can then be used in virtual scenes or hybrid/natural synthetic videos. To check from a certain view if an object is protected or not, the 2D contour that defines the boundary of the object's projection is extracted, sampled and the watermark's presence is detected.

Figure 4.22 summarizes all the techniques described in this chapter.

4.5 Concluding Remarks

During the last few years, watermarking of 3D objects has attracted a considerable amount of interest within the watermarking community, although not as much as watermarking of other types of multimedia data like images, audio or video. Although 3D object watermarking algorithms have many things in common with algorithms developed for other types of media, the particularities of 3D data, e.g. the fact that no natural ordering (and thus no globally accepted traversal scheme) can be devised for points in the 3D space or the fact that, unlike other media, watermark imperceptibility should not be judged directly on the 3D object but rather on its projections (images,

Algorithm	Representation format	Extraction mode	Embedding domain	Robust to remeshing	Remarks
1	Mesh	Blind	Computer data	No	Reordering of vertices or triangles
2	NURBS	Non-blind	Computer data	—	Reparameterization
3	CSG	Blind	Topological data	—	
4	Mesh	Blind	Topological data	No	Triangle Strip Peeling Symbol
5	Mesh	Blind	Topological data	Yes	Polygonal Stencil Pattern
6	Mesh	Blind	Topological data	Yes	Mesh Density Pattern
7	Mesh	Blind	Topological data	No	Triangle subdivision
8	Mesh	Blind	Geometrical data	No	Vertex displacement
9	Mesh	Non-blind	Geometrical data	Yes	Vertex displacement
10	Mesh	Blind	Geometrical data	Yes	Vertex displacement
11	Mesh	Blind	Geometrical data	No	Fragile watermark
12	Mesh	Blind	Geometrical data	Yes	Vertex Flood
13	Mesh	Non-blind	Geometrical data	Yes	Vertex displacement
14	Mesh	Blind	Geometrical data	No	Triangle Flood
15	Mesh	Blind	Geometrical data	Yes	Triangle Similarity Quadruple
16	Mesh	Blind	Geometrical data	No	Tetrahedral Volume Ratio
17	Mesh	Blind	Geometrical data	No	Triangle Strip Peeling Symbol
18	Mesh	Blind	Geometrical data	No	Length of normals
19	Mesh	Blind	Geometrical data	Yes	Repartition of normals
20	Mesh	Blind	Geometrical data	Yes	Repartition of normals
21	Mesh	Blind	Geometrical data	No	Cylindrical depth map
22	Mesh	Blind	Geometrical data	No	Principal component analysis
23	Mesh	Blind	Geometrical data	Yes	Sectional Principal Object Axis
24	Mesh	Blind	Geometrical data	No	progressive transmission

Figure 4.22   List of 3D watermarking algorithms

Algorithm	Representation format	Extraction mode	Embedding domain	Robust to remeshing	Remarks
25	Mesh	Non-blind	Geometrical data	No	Wavelet decomposition
26	Mesh	Blind	Geometrical data	Yes	Wavelet decomposition
27	Mesh	Non-blind	Geometrical data	Yes	Wavelet decomposition
28	Mesh	Non-blind	Geometrical data	Yes	Multiresolution representation
29	Mesh	Non-blind	Geometrical data	Yes	Spherical wavelet transform
30	Mesh	Non-blind	Geometrical data	Yes	Spectral decomposition
31	Mesh	Non-blind	Geometrical data	Yes	Spectral decomposition
32	Mesh	Blind	Geometrical data	Yes	Spectral decomposition
33	Mesh	Non-blind	Geometrical data	Yes	Spectral decomposition
34	Mesh	Non-blind	Geometrical data	No	Singular spectrum analysis
35	Mesh	Blind	Geometrical data	No	Generalized Radon transform
36	Mesh	Non-blind	Geometrical data	Yes	Discrete cosine transform
37	NURBS	Blind	Geometrical data	—	
38	NURBS	Blind	Geometrical data	—	Orthographic depth map
39	NURBS	Blind	Geometrical data	—	Spread spectrum
40	Voxels	Blind	Geometrical data	—	
41	Point set	Non-blind	Geometrical data	—	
42	—	—	—	—	Other attributes
43	—	—	—	—	Animation parameter
44	—	Non-blind	—	—	Asymmetrical approach
45	Mesh	Non-blind	Geometrical data	—	Asymmetrical approach

Figure 4.22 (*continued*)

videos) on the plane, make 3D watermarking a rather distinct and definitely interesting problem. Despite the significant amount of work conducted so far in this area (especially on mesh models), numerous open issues exist and the problem is far from being considered solved. Obviously, it is rather optimistic to believe that a 3D object watermarking technique that is robust to all envisioned attacks (including the ones that will be designed in the future) will be introduced in the near future. However, watermarking can indeed be a powerful digital rights management tool provided that the devised techniques are constructed bearing in mind the needs and challenges of specific applications and environments. The combination of watermarking with other related techniques like encryption or perceptual hashing is also a promising direction for the successful application of 3D object watermarking in real-world scenarios. The introduction of universally accepted benchmarking procedures, test sets and performance metrics can also help in this regard.

4.6   Summary

Due to the expansion of the Internet and the multiplication of high-flow networks, multimedia databases now contain more and more 3D objects. Real applications show that 3D watermarking can be useful for several purposes, security-based ones being the most prominent. For example, users might like to check if the use of a given object is legal or not, and to access additional information concerning the object (e.g. for authentication or indexing), the owner (copyright), or even the buyer (e.g. for traitor tracing).

The subject of 3D watermarking consists of hiding in an invisible way information in a 3D object. This information can then be recovered at any time even if the 3D object was altered by one or more non-destructive attacks, either malicious or not.

By analogy with still image watermarking, which was historically used in the first types of documents investigated for watermarking, 3D watermarking techniques can be classified in two major categories with respect to the information conveyed by the watermark. We can distinguish *zero-bit* techniques, where one can only verify whether the cover is watermarked or not, and *multiple-bit* techniques, where if the 3D object is watermarked, the embedded message should be decoded. The 3D watermarking algorithm can also be classified with respect to the detection techniques into *blind* (we need only the watermark key for extraction) and *non-blind* (we need the watermark key as well as the original 3D object for extraction) techniques.

To insert information on the 3D object to get it watermarked, three techniques can be distinguished according to the embedding space. We can encode information by modifying the organization of the data in the computer file associated with the 3D object; typically authors propose to modify in the computer file the order of triangles within a list of triangles or the order of triplets of vertices forming a given triangle. Information can also be encoded using the topology of the 3D object, i.e. the connectivity of the mesh. Finally, 3D watermarking can operate on the geometric data of the object by slightly modifying the positions of vertices.

It is worth mentioning that watermarking 3D objects involves a trade off between *capacity* (the amount of information that can be hidden), *visibility* (the visual degradation of the 3D object due to watermarking) and *robustness* (the ability to recover the watermark even if the 3D watermarked object has been manipulated).

4.7   Questions and Problems

4.7.1   Watermarking Questions

1. What are the advantages of watermarking in different domains, e.g. DCT versus DWT (Discrete Wavelet Transform)?
2. Describe the characteristics and differences of the following watermark categories:

   (a) robust watermarks;
   (b) fingerprints;
   (c) fragile watermarks.

3. What are the differences between:

   (a) de-synchronization and removal attacks;
   (b) blind and non-blind watermark extraction?

4. Give an example of a 3D watermarking algorithm that operates in:

   (a) computer data (organization of the data within the file);
   (b) topological data;
   (c) geometrical data.

5. Explain the trade off between capacity, visibility and robustness in the case of 3D watermarking.

6. What are the possible applications of 3D object watermarking?
7. Give some examples of manipulations and attacks on 3D objects.

4.7.2 Watermarking Problems

A polyhedron can be defined by a set of vertices and a set of polygonal faces constructed by connecting these vertices. Such data can be represented as a computer file formatted as follows:

$$N_p$$
$$X_1, Y_1, Z_1$$
$$\vdots$$
$$X_i, Y_i, Z_i$$
$$\vdots$$
$$X_{N_p}, Y_{N_p}, Z_{N_p}$$
$$N_f$$
$$N_{p_1}, n_{1,1}, \ldots, n_{1,k}, \ldots, n_{1,N_{p_1}}$$
$$\vdots$$
$$N_{p_j}, n_{j,1}, \ldots, n_{j,k}, \ldots, n_{j,N_{p_j}}$$
$$\vdots$$
$$N_{p_{Nf}}, n_{N_f,1}, \ldots, n_{N_f,k}, \ldots, n_{N_f,N_{p_{N_f}}}$$

- N_p is the number of vertices, and is followed by an ordered list of N_p triplets of coordinates.
- The triplet X_i, Y_i, Z_i, $1 \leq i \leq N_p$, represents the 3D coordinates of the ith vertex. In our format, these coordinates are integer values.
- N_f is the number of polygonal faces of the polyhedron, and is followed by N_f lines, each describing a different face.
- A line of the form N_{p_j}, $n_{j,1}, \ldots, n_{j,k}, \ldots, n_{j,N_{p_j}}$, $1 \leq j \leq N_f$, describes the jth polygonal face. N_{p_j} is the number of vertices of the face and it is followed by an ordered list of the indices ($n_{j,1}$ to $n_{j,N_{p_j}}$) of the vertices that define the boundary of this face.

As an illustration, the cube of Figure 4.23 can be represented by the following file:

$$8$$
$$-500, -500, 500$$

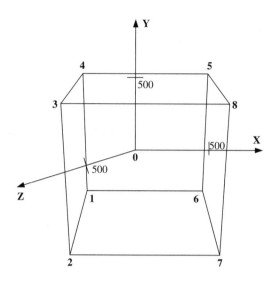

Figure 4.23 Example of a cube

$$-500, -500, 500$$
$$-500, 500, 500$$
$$-500, 500, -500$$
$$500, 500, -500$$
$$500, 500, 500$$
$$500, -500, 500$$
$$500, -500, -500$$
$$6$$
$$4, 5, 6, 7, 8$$
$$4, 1, 2, 3, 4$$
$$4, 2, 3, 6, 7$$
$$4, 4, 5, 6, 3$$
$$4, 1, 4, 5, 8$$
$$4, 1, 2, 7, 8$$

1. We want to hide information in the file that represents a polyhedron in such a way that the shape of the polyhedron is modified as little as possible. To do so, we change the LSB (Least Significant Bit) of the numbers X_i, Y_i, Z_i.

(a) In the general case, what amount of data (i.e. how many bits) can we hide with this method?

(b) In the case of the cube given as an example, does the insertion of data in the LSB significantly modify its shape? (Compute an order of magnitude of the ratio between the maximal displacement of a vertex and the size of the cube.)

(c) To increase the amount of hidden information while preserving the shape of the object, could we modify the LSB of N_p, N_f, N_{p_j}, $n_{j,k}$ too?

2. Now we assume that after having watermarked an object with the previous method, the file is modified by an attacker who arbitrarily renumbers the vertices. Thus, a permutation s is applied to the vertices: vertex number i becomes vertex number $s(i)$. In practice, this means that the line describing vertex number i is moved to the line number $s(i)$ in the list of vertices, and that i is replaced by $s(i)$ whenever it appears among the $n_{j,k}$ that describe the vertices of a face. This renumbering of the vertices does not change the shape of the object.

(a) Blind extraction after renumbering attack: Is it possible to recover the information hidden in the file after the vertices have been renumbered by an unknown permutation, and knowing only the watermarked and renumbered file?

(b) Non-blind extraction after renumbering attack: Is it possible to recover the hidden information if, along with the watermarked and renumbered file (see previous question), we also know the original unwatermarked file?

3. Suppose that we use another watermarking scheme that encodes hidden information by modifying the order of the list of faces, an operation that does not change the shape of the described object. To do this, we first choose a reference order of the faces by sorting them, more precisely by sorting the lists N_{p_j}, $n_{j,1}$, ..., $n_{j,k}$, ..., $n_{j,N_{p_j}}$ $(1 \leq j \leq N_f)$ in lexicographic order. By doing so for the cube example we obtain:

$$4, 1, 2, 3, 4$$
$$4, 1, 2, 7, 8$$
$$4, 1, 4, 5, 8$$
$$4, 2, 3, 6, 7$$
$$4, 4, 5, 6, 3$$
$$4, 5, 6, 7, 8$$

Then, we number the N_f! possible permutations of this list of faces choosing an arbitrary convention, and we encode a number N in the file that represents the 3D object by ordering the list of faces using the Nth permutation of the reference order.

(a) How do we extract the hidden information (the number N) from the watermarked file? (Describe only the general idea in a few lines, not the complete algorithm.)

(b) In the general case, what is the capacity in bits of this watermarking algorithm?

(c) Is it possible to recover the hidden information in blind mode (i.e. knowing only the watermarked file) if the vertices have been renumbered in an unknown way by an attacker after watermarking?

(d) What is your answer to the previous question in the case of non-blind detection, i.e. when both the attacked watermarked file and the original (unwatermarked) file are available?

4. We use the same watermarking principle as in Question 3 but this time we add a preliminary step: firstly, we renumber the vertices so that they are ordered in lexicographic order with respect to their coordinates X_i, Y_i, Z_i, and then we watermark the file by modifying the order of the faces as before.

(a) Assume that an attacker renumbers the vertices after watermarking. How is it possible to recover the hidden information in blind mode (from the attacked watermarked file only)?

(b) Is this watermarking algorithm robust to noise addition on the coordinates of the vertices?

References

Alface P and Macq B 2006 Blind watermarking of 3D meshes using robust feature points detection. *IEEE International Conference on Image Processing*, Piscataway, New Jersey.

Ashourian M and Enteshary R 2003 A new masking method for spatial domain watermarking of three dimensional triangle meshes. *TENCON 2003. Conference on Convergent Technologies for Asia–Pacific Region*, vol. 1, pp. 428–431.

Barni M, Bartolini F, Cappellini V, Corsini M and Garzelli A 2004 Digital watermarking of 3D meshes. In *Mathematics of Data/Image Coding, Compression, and Encryption VI, with Applications*. ed. Schmalz, MS. *Proceedings of the SPIE*, **5208**, 68–79.

Benedens O 1999a Geometry-based watermarking of 3D models. *IEEE Computer Graphics and Applications* **19**(1), 46–55.

Benedens O 1999b Two high capacity methods for embedding public watermarks into 3D polygonal models. *ACM Multimedia and Security Workshop*, Orlando, Florida, pp. 95–99.

Benedens O 1999c Watermarking of 3D polygon based models with robustness against mesh simplification. *SPIE Security and Watermarking of Multimedia Content*, pp. 329–340.

Benedens O 2000 Affine invariant watermarks for 3D polygonal and NURBS based models. *Third Information Security Workshop*, Wollongong, Australia.

Benedens O and Busch C 2000 Towards blind detection of robust watermarks in polygonal models. *Computer Graphics Forum*, **19**(3), 199–208.

Bennour J and Dugelay JL 2006 Protection of 3D object visual representations *Multimedia and Expo, 2006 IEEE International Conference, Toronto*, pp. 1113–1116.

Bors AG 2004a Blind watermarking of 3D shapes using localized constraints. *IEEE 2nd International Symposium on 3D Data Processing, Visualization and Transmission*, pp. 242–249.

Bors AG 2004b Watermarking 3D shapes using local moments. *IEEE International Conference on Image Processing*, Singapore, vol. 1, pp. 729–732.

Bors AG 2006 Watermarking mesh-based representations of 3-D objects using local moments. *IEEE Transactions on Image Processing* **15**(3), 687–701.

Carr JC, Beatson RK, Cherrie JB, Mitchell TJ, Fright WR, McCallum BC and Evans TR 2001 Reconstruction and representation of 3D objects with radial basis functions. In *SIGGRAPH 2001, Computer Graphics Proceedings* (ed. Fiume E), pp. 67–76. ACM Press/ACM SIGGRAPH.

Cayre F and Macq B 2003 Data hiding on 3D triangle meshes. *IEEE Transactions on Signal Processing, Special Issue on Data Hiding and Secure Delivery of Multimedia Content* **51**(4), 939–949.

Cayre F, Rondao-Alface P, Schmitt F, Macq B and Matre H 2003 Application of spectral decomposition to compression and watermarking of 3D triangle mesh geometry. *Image Communications, Special Issue on Image Security* **18**(4), 309–319.

Cox I, Kilian J, Leighton T and Shamoon T 1997 Secure spread spectrum watermarking for multimedia. *IEEE Transactions on Image Processing* **6**(12), 1673–1687.

Cox IJ, Miller ML and Bloom JA 2002 *Digital Watermarking*. Morgan Kaufmann.

Daras P, Zarpalas D, Tzovaras D and Strintzis M 2004 3D model watermarking for indexing using the generalized Radon transform. *IEEE. 2nd International Symposium on 3D Data Processing, Visualization and Transmission*, pp. 102–109.

Fornaro C and Sanna A 2000 Public key watermarking for authentication of CSG models. *Computer Aided Design* **32**(12), 727–735.

Galka A 2001 *Topics in Nonlinear Time Series Analysis*. World Scientific.

Garcia E and Dugelay JL 2003 Texture-based watermarking of 3D video objects. *IEEE Transactions on Circuits and Systems for Video Technology* **13**(8), 853–866.

Gelasca D, Ebrahimi T, Corsini M and Barni M 2005 Objective evaluation of the perceptual quality of 3D watermarking. *IEEE International Conference on Image Processing (ICIP)*.

Golyandina N, Nekrutkin V and Zhigljavsky A 2001 *Analysis of Time Series Structure: SSA and Related Techniques*. Chapman & Hall/CRC Press.

Harte T and Bors AG 2002 Watermarking 3D models. *IEEE International Conference on Image Processing*, Rochester, New York.

Hartung F, Eisert P and Girod B 1998 Digital watermarking of MPEG-4 facial animation parameters. *Computers & Graphics* **22**(4), 425–435.

Hoppe H 1996 Progressive meshes. *ACM SIGGRAPH*, pp. 99–108.

Ichikawa S, Chiyama H and Akabane K 2002 Redundancy in 3D polygon models and its application to digital signature. *Journal of WSCG* **10**(1), 225–232.

Jeonghee J, Lee SK and Ho YS 2003 A three-dimensional watermarking algorithm using the DCT transform of triangle strips. *Digital Watermarking Second International Workshop, IWDW*, Berlin.

Jin JQ, Dai MY, Bao HJ and Peng QS 2004 Watermarking on 3D mesh based on spherical wavelet transform. *Journal of Zhejiang University SCIENCE* pp. 251–258.

Kalivas A, Tefas A and Pitas I 2003 Watermarking of 3D models using principal component analysis *Maltimedia and Expo, ICME'03, Proceedings*, pp. 637–640.

Kanai S, Date H and Kishinami T 1998 Digital watermarking for 3D polygons using multiresolution wavelet decomposition. *6th IFIP 5.2 International Workshop on Geometric Modeling: Fundamentals and Applications (GEO-6)*, Tokyo, pp. 296–307.

Katzenbeisser S and Petitcolas FAP 2000 *Information Hiding Techniques for Steganography and Digital Watermarking*. Artech House.

Koh B and Chen T 1999 Progressive browsing of 3D models. *IEEE Workshop on Multimedia Signal Processing*, Copenhagen.

Kwon K, Kwon S, Lee S, Kim T and Lee K 2003 Watermarking for 3D polygonal meshes using normal vector distributions of each patch. *IEEE International Conference on Image Processing*, pp. II: 499–502.

Lee JJ, Cho NI and Kim JW 2002 Watermarking for 3D NURBS graphic data. *IEEE International Workshop on MultiMedia Signal Processing*.

Lindstrom P and Turk G 2000 Image-driven simplification. *ACM Transactions on Graphics* **19**(3), 204–241.

Louizis G, Tefas A and Pitas I 2002 Copyright protection of 3D images using watermarks of specific spatial structure. *IEEE Multimedia and Expo ICME* **2**, 557–560.

Mao X, Shiba M and Imamiya A 2001 Watermarking 3D geometric models through triangle subdivision. In *Proceedings of the SPIE, Security and Watermarking of Multimedia Contents III* (ed. Wong PW and Delp EJ), **4314**, 253–260.

Maret Y and Ebrahimi T 2004 Data hiding on 3D polygonal meshes. *MM&Sec'04: Proceedings of the 2004 workshop on Multimedia and Security*, pp. 68–74. ACM Press.

Mitrea M, Zaharia T and Preteux F 2004 Spread spectrum robust watermarking for NURBS surfaces. *WSEAS Transactions on Communications* **3**(2), 734–740.

Murotani K and Sugihara K 2003 Watermarking 3D polygonal meshes using the singular spectrum analysis. *IMA International Conference on the Mathematics of Surfaces*, pp. 19–24.

Murotani K and Sugihara K 2004 Generalized SSA and its applications to watermarking 3D polygonal meshes. In *METR*, pp. 1–23.

Ohbuchi R and Masuda H n.d. Managing CAD data as a multimedia data type using digital watermarking.

Ohbuchi R, Masuda H and Aono M 1997 Watermarking three-dimensional polygonal models. *ACM Multimedia*, Seattle, Washington, pp. 261–272.

Ohbuchi R, Masuda H and Aono M 1998a Geometrical and non-geometrical targets for data embedding in three-dimensional polygonal models. *Computer Communications* **21**, 1344–1354.

Ohbuchi R, Masuda H and Aono M 1998b Watermarking multiple object types in three-dimensional models. *Multimedia and Security Workshop at ACM Multimedia*, Bristol, UK.

Ohbuchi R, Masuda H and Aono M 1999 A shape-preserving data embedding algorithm for NURBS curves and surfaces. *Computer Graphics International*, pp. 170–177.

Ohbuchi R, Mukaiyama A and Takahashi S 2002 A frequency-domain approach to watermarking 3D shapes. *Eurographics 2002*, Saarbrucken.

Ohbuchi R, Mukaiyama A and Takahashi J 2004 Watermarking a 3D shape model defined as a point set. *International Conference on Cyberworlds*.

Ohtake Y, Belyaev A and Seidel H 2004 3D scattered data approximation with adaptive compactly supported radial basis functions. *SMI'04: Proceedings of Shape Modeling International 2004*, Washington, DC, pp. 31–39.

Praun E, Hoppe H and Finkelstein A 1999 Robust mesh watermarking. *ACM SIGGRAPH*, Los Angeles, California, pp. 49–56.

Rogowitz B and Rushmeier H 2001 Are image quality metrics adequate to evaluate the quality of geometric objects? *Proceedings of the SPIE, Human Vision and Electronic Imaging VI*, **4299** 340–348.

Song HS and Cho NI 2004 Digital watermarking of 3D geometry. *Intelligent Signal Processing and Communication Systems*, pp. 272–277.

Song HS, Cho NI and Kim JW 2002 Robust watermarking of 3D mesh models. *IEEE International Workshop on MultiMedia Signal Processing*.

Taubin G, Zhang T and Golub G 1996 Optimal surface smoothing as filter design. Technical Report RC-2040, IBM.

Tefas A and Pitas I 2001a Robust spatial image watermarking using progressive detection.

Tefas A and Pitas I 2001b Robust spatial image watermarking using progressive detection. *Proceedings of the IEEE International Conference on Acoustics, Speech, and Signal Processing, vol. 3.*

Tefas A, Nikolaidis N and Pitas I 2005 *Watermarking Techniques for Image Authentication and Copyright Protection*, 2nd edition, Elsevier Academic Press.

Uccheddu F, Corsini M and Barni M 2004 Wavelet based blind watermarking of 3D models. *Multimedia and Security Workshop*, pp. 143–154.

Valette S, Kim Y, Jung H, Magnin I and Prost R 1999 A multiresolution wavelet scheme for irregularly subdivided 3D triangular mesh, *IEEE International Conference on Image Processing (ICIP'99)* pp. I:171–174.

Wagner MG 2000 Robust watermarking of polygonal meshes. *Geometric Modeling and Processing*, Hong Kong.

Wu J and Kobbelt L 2005 Efficient spectral watermarking of large meshes with orthogonal basis functions. *The Visual Computer* **21**(8–10), 848–857.

Yeo BL and Yeung MM 1999 Watermarking 3D objects for verification. *IEEE Computer Graphics and Applications* pp. 36–45.

Yeung M and Yeo BL 1998 Fragile watermarking of three-dimensional objects. *IEEE International Conference on Image Processing*, vol. 2, pp. 442–446.

Yin K, Pan Z, Shi J and Zhang D 2001 Robust mesh watermarking based on multiresolution processing. *Computers & Graphics* **25**, 409–420.

Yu Z, Ip H and Kwok L 2003 A robust watermarking scheme for 3D triangular mesh models. *Pattern Recognition* **36**(11), 2603–2614.

Zafeiriou S, Tefas A and Pitas I 2005 Blind robust watermarking schemes for copyright protection of 3D mesh objects. *IEEE Transactions on Visualization and Computer Graphics* **11**(5), 596–607.

Zhi-qiang Y, Ip HHS and Kowk LF 2003 Robust watermarking of 3D polygonal models based on vertice scrambling. *IEEE CGI Computer Graphics International*, p. 254.

Conclusion

Three-dimensional processing is an emerging and growing need in the multimedia domain. There are numerous application domains that increasingly use 3D objects, e.g. biometrics based on 3D face recognition or 3D object creation, visualization and exchange in computed-assisted design, or 3D acquisition and processing for control quality and biomedical needs. Thus 3D processing is no longer limited to graphics and synthesis but also includes multidimensional signal processing tools.

As for images, video and audio, the techniques of compression, indexing and watermarking described in this book are of interest for 3D objects. Working on any of these functionalities requires minimum background knowledge in information theory, multidimensional signal processing, data analysis tools and 3D modeling. This book is organized as follows: an introduction; basic knowledge on 3D modeling, compression, indexing and watermarking; and the present short conclusion. One can readily note that the state of the art in the three functionalities is unbalanced. There exists much more work on compression as compared to indexing and watermarking. Also, compression of 3D objects is now included in some standards, e.g. MPEG-4, whereas watermarking of 3D objects, still immature, has not yet really been taken into consideration.

Even if there are many ways to describe a 3D object (from both mathematical and software points of view), as seen in the first chapter, polygonal meshes are by far the most popular representation. In particular, 99 % of the existing algorithms dealing with compression, indexing or watermarking apply on such meshes. The predominance of this model is mainly due to its algebraic simplicity.

One of the main difficulties encountered in 3D model processing is the question of distortion evaluation. Indeed, compression or watermarking introduce slight degradations to the shape of the model and it is much more difficult to evaluate the corresponding visual distortion than for 2D images. In particular, it is not easy to find an exact correspondence between the vertices of two meshes. However, some efficient distortion metrics are progressively emerging for 3D models.

In the chapter on compression, we have presented existing 3D compression techniques and more particularly polygonal mesh compression algorithms. The research community has been particularly active in the latter field for the last 10 years.

Mono-resolution techniques now provide compression rates between 10 and 16 bits per vertex. The initial techniques focused more on connectivity, but recent approaches give priority to the geometry that represents the major part of the data.

Concerning multi resolution techniques, whereas the constraint of progressiveness involved a strong reduction in the effectiveness of the first approaches, recent algorithms have reached the compression rates obtained by mono-resolution ones.

Most effective algorithms today rather focus on compressing the shape of the object than the mesh itself. They carefully remesh the object (regularly or according to feature lines, for instance) so as to improve the performances of further compression schemes. Indeed, the quality of the mesh represents a critical issue since it drives the efficiency of the compression. In the same manner, another efficient way to obtain high compression rates is to change the whole representation of the object using approximation schemes. For example, approximating a dense polygonal mesh with an implicit or subdivision surface, or with a set of geometric primitives like superellipsoids, leads to extremely high compression rates since these models are far more compact. However, these algorithms are quite complex to handle. In our view, there are very challenging issues on shape modeling and coding, especially for very low bit rates preserving the visual quality. These areas are under investigation by several research teams around the world.

In the chapter on indexing, we have presented the foundations and the methodology to build a 3D model search engine. We have described a variety of invariant descriptors for 3D objects. We have shown the importance of having a rigorous methodology and criteria to compare a new 3D retrieval algorithm to state-of-the-art methods.

Thanks to the availability of technologies for their acquisition, 3D search engines are being employed in a wide range of application domains, including medicine, computer-aided design and engineering, cultural heritage, videogames and entertainment applications, and so on. Important open questions remain in the search for the content-based description and retrieval of 3D objects:

- How can the efficiency of 3D search systems be improved when the dimension of the descriptors becomes high?
- How can these techniques be generalized for huge databases in order to ensure real-time or at least on line retrieval?
- How can partial query and retrieval be managed? One would then need 3D object segmentation, which is always an open and difficult problem. Note that the segmentation also leads to an increase in the database size with parts of 3D objects instead of the whole objects themselves.
- Should the current methods focus mainly on geometric aspects of 3D models? Color and texture are present in many databases, such as cultural heritage databases, and should be taken into account for future research as joint descriptors.

As shown in the chapter on watermarking, during the last few years, watermarking of 3D objects has attracted a considerable amount of interest within the research community, but not as much as watermarking of other types of multimedia data like images, audio or video.

Although 3D object watermarking algorithms have many things in common with those algorithms developed for other media types, the well-known 2D schemes could not and should not be generalized directly for 3D schemes. Indeed, 3D data have their own particularities: the fact that no natural ordering can be devised for points in the 3D space; the fact that, unlike other media, watermark imperceptibility can be judged on the 3D object and also on its projections (images, videos) in the plane. These make 3D watermarking a rather distinct and definitely interesting problem. Despite the significant amount of work conducted so far in this area (especially on mesh models), numerous open issues exist and the problem is far from being considered solved. Clearly, it is rather optimistic to believe that a 3D object watermarking technique that is robust to all envisioned attacks (including those yet to be designed) will be introduced in the near future. However, watermarking can indeed be a powerful digital rights management tool provided that the devised techniques are constructed bearing in mind the needs and challenges of specific applications and environments. The combination of watermarking and other related techniques like encryption or perceptual hashing or joint source channel vision constitutes a promising direction for the successful application of 3D object watermarking in real-world scenarios. The introduction of universally accepted benchmarking procedures, test sets and performance metrics can also help here.

As described in this book, even if there is still much room for improvement, many of the 2D processing tools available, such as mathematical operators or wavelet decomposition, are now proposed in 3D. New processing tools are needed in order to better understand, analyze and model the complex nature of 3D objects. It is important to bear in mind that advances in 3D processing also depend on the availability of data that are mandatory for tests and significant evaluations: that is to say, advances in technologies (i.e. specialized scanners) and the existence of databases.

If we consider the specific case of face recognition as a significant example of the use of 3D, the situation in terms of pick-up systems and databases is still very immature compared to classical images.

One can also note that instead of using customized video cameras or scanners, some works proposed to build 3D models from video or a pair of images (e.g. frontal and profile). Nevertheless, this method imposed some technical constraints on the acquisition and some restrictions on the resolution that were often unsuitable for real applications.

On the contrary, for facial pictures or even facial videos, 3D face databases exist but are very rare. The lack of databases is a critical issue because they are mandatory for evaluating performances associated with new 3D algorithms. Moreover, for open problems in 3D processing, the objective and subjective evaluation of quality is another critical issue. It is well known that, even in 2D processing, measures such as PSNR (Peak Signal-to-Noise Ratio) are imperfect, a problem also existing in 3D processing. Until now, there have been neither objective protocols as in audio and video to evaluate the impact of lossy compression (e.g. CCITT protocols), nor benchmarking tools as in image watermarking (e.g. Stirmark) to evaluate the robustness of a watermark, for example. Cost and complexity of 3D scanners could be handicaps to the dissemination of 3D technologies and associated software.

In terms of perspective, new work in 3D processing could consist of studying functionalities (i.e. compression, indexing and watermarking) jointly and not separately. It is well known that lossy compression and watermarking operations may conflict with each other. Indexing may be useful to recover watermarks by (pre)selecting a subset of objects inside a large database, etc. There are already some technical connections between the three domains, in particular the use of hierarchical and multiresolution approaches when processing 3D objects represented by meshes.

Index